The Cambridge Manuals of Science and Literature

THE FERTILITY OF THE SOIL

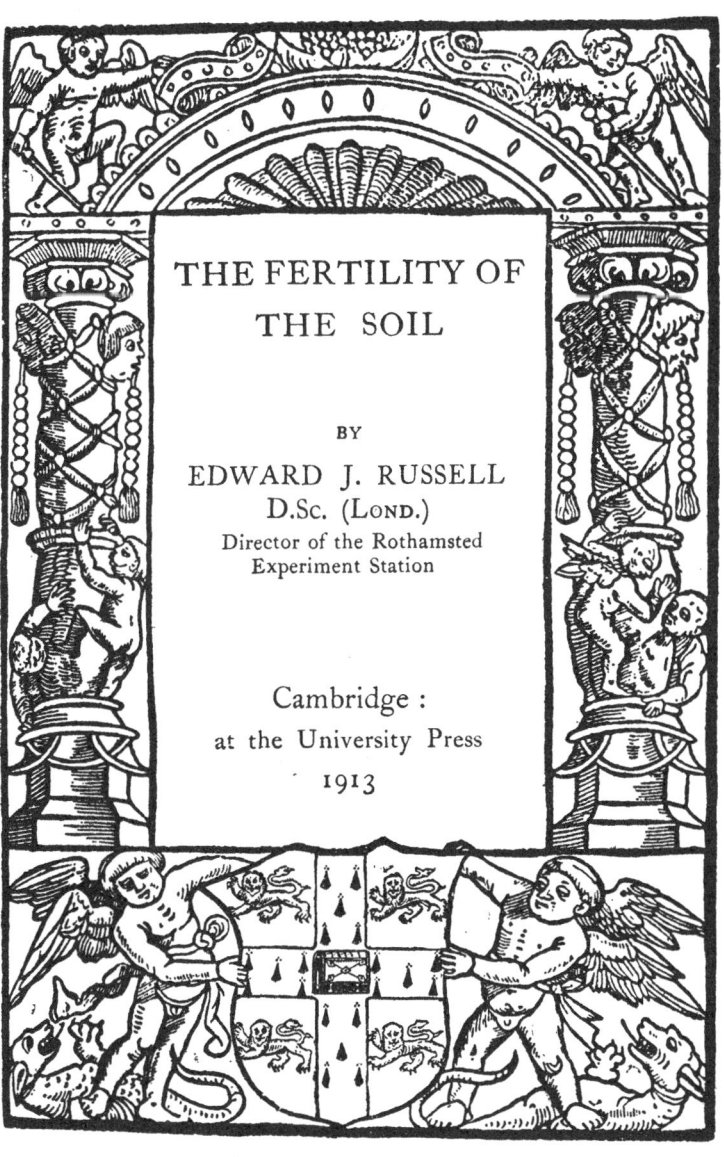

THE FERTILITY OF THE SOIL

BY

EDWARD J. RUSSELL

D.Sc. (Lond.)

Director of the Rothamsted
Experiment Station

Cambridge :

at the University Press

1913

CAMBRIDGE UNIVERSITY PRESS
Cambridge, New York, Melbourne, Madrid, Cape Town,
Singapore, São Paulo, Delhi, Tokyo, Mexico City

Cambridge University Press
The Edinburgh Building, Cambridge CB2 8RU, UK

Published in the United States of America by
Cambridge University Press, New York

www.cambridge.org
Information on this title: www.cambridge.org/9781107401761

© Cambridge University Press 1913

First published 1913
First paperback edition 2011

A catalogue record for this publication is available from the British Library

ISBN 978-1-107-40176-1 Paperback

Cambridge University Press has no responsibility for the persistence or
accuracy of URLs for external or third-party internet websites referred to in
this publication, and does not guarantee that any content on such websites is,
or will remain, accurate or appropriate.

*With the exception of the coat of arms
at the foot, the design on the title page is a
reproduction of one used by the earliest known
Cambridge printer, John Siberch, 1521*

PREFACE

THE following pages contain the substance of talks, lectures and other discourses delivered before all sorts and conditions of men and women and in all kinds of meeting places. Sometimes the listeners were labourers and allotment holders gathered by the schoolmaster at the close of the day in the biggest room of the village inn, or in the adult school on Sunday morning; sometimes they were the more polite but not more interested cultivators of suburban gardens; others were farmers assembled in the market town on market day; others again were professed and serious students of agricultural science. But all had this in common: that they were really keenly interested in the soil they were cultivating, and wanted to know something more about it. It is for such readers that this little book is intended.

E. J. R.

HARPENDEN,
July 1913.

CONTENTS

LIST OF ILLUSTRATIONS

CHAPTER I

THE NATURAL HISTORY OF THE SOIL

To those who have never thought about the matter the study of the soil may seem very trivial; it has neither the glory of the celestial nor the glamour of the unfamiliar; it is associated with such unintellectual and mundane concerns as food production, and has no place in our ordinary conception of a refined and liberal education.

But the soil has not always been looked upon as commonplace. In the mythology of Greece it held a very dignified position, the Goddess Gaea being the mother of mankind and the bounteous provider of food. Right through into much later times this idea of the kindly Mother Earth can be traced, and even to-day the reflective gardener takes more than a utilitarian interest in his soil. And the light of Science more than justifies this interest, for it has shown that the soil is far more wonderful than any human mind had ever pictured it.

In trying to trace out the history of a lump of soil we must go back to those remote times when the earth was first cool enough to allow a solid crust to form. When the water began to fall some of it soaked into the chinks and crannies of this crust, and by its expansion and contraction with change of temperature caused fragments to split off from the main rock. Other agencies were also effecting the same end, and in course of time a great quantity of this disintegrated rock matter was formed. The particles did not remain where they were, but were carried by wind or water into the valleys and streams and many found their way to the bottom of the sea. Here they mingled with the residues of plants and animals, and the whole mass became consolidated. Later on, when earth movements changed the course of the waters and the old sea became dry land, this consolidated material appeared as new rock and went again through the processes of disintegration and erosion. For a long period the surface of England north of the Thames, and of Canada and the northern part of the United States, was covered with ice which pounded up and carried away many of the particles, depositing them again when it melted. When the particles were lying on the dry land they were subject to the constant washing of the rain, the oxidising effect of the atmosphere, and the shattering effect of changing temperature, processes collectively known as weathering. Ever since the particles first split off from the original rock they have been exposed to these continuous disintegrating processes. But the action of these processes is exceedingly slow, or the particles would have disappeared entirely, or have become reduced to an impalpable dust. The fact that they survived proves them to be very resistant and indicates that they are not likely to undergo any appreciable change during the short period of time that interests the agriculturalist. Over longer periods of time, however, the different particles show different degrees of resistance, the most resistant being the grains of quartz and the least resistant the more complex combinations of silica and oxides of iron, aluminium, potassium and other metals. The latter have therefore suffered more than the quartz and have been reduced to much finer dimensions. Thus if a soil is separated out by mechanical analysis into portions, the particles of which fall within certain definite limits of size, it will be found that the coarsest particles of all—the stones and gravel—represent complex rock material, the coarse particles of the fine earth (the so-called coarse sand, fine sand and silt) are practically pure silica, while the finest particles (the clay and to a less extent the fine silt) contain not only silica, but oxides of iron, aluminium and of other metals as well. Further, the top 8 inches of soil that has been exposed to weathering processes for very long periods of time contains practically as much coarse material (silica) as the subsoil which has been shielded from these actions, but it contains markedly less of the finest material.

These particles constitute the chief portion of the soil and may be regarded as the framework round which the soil is built. They show, however, certain differences which are of fundamental importance to the subject. The sands and fine silt, being formed of the silica, are chemically inert and practically unalterable under natural conditions except that they may very slowly be reduced in size by weathering processes. The finer material, on the other hand, is chemically active and may not only undergo chemical changes, but may enter into reaction with various substances; it is not a single definite compound like the silica, however, and cannot be represented by any chemical formula. It possesses other properties that mark it off very sharply from the coarse material. It absorbs a considerable amount of water, swelling up very much during the process; conversely when it dries it shrinks a good deal. In its wet state it is very sticky, when dry it is very hard. It undergoes a remarkable change after treatment with traces of acids or of salts, notably calcium carbonate, and becomes less sticky and more easily crumbled. All these properties are readily observed in a clay field where much of this fine material is present: the persistent wetness of the soil, its stickiness, the large gaping cracks that form during its contraction on drying and the hard clots that result, its marked alteration after treatment with lime, are all manifestations of the special properties of the fine material. These properties are characteristic of the jelly-like condition, technically known as the colloidal state, into which many substances can be brought.

These properties are not shown by the coarse material; a sandy soil (in which the coarse particles predominate and the fine particles do not form more than $5-10\%$ of the whole) has no great power of absorbing water and therefore readily dries, it is not sticky, does not shrink on drying or form hard clods, and undergoes no obvious physical change after treatment with lime.

It must not be supposed that any hard and fast line can be drawn between the coarse sand material and the fine clay. One shades off imperceptibly into the other, and so gradual is the transition that the special name silt is used to designate the intervening material. The lines separating sand from silt on the one hand and silt from clay on the other are purely conventional and are agreed upon by soil chemists in each country, but unfortunately no two countries accept quite the same definitions. In Great Britain clay is defined as any material the particles of which are less than ·002 mm. in diameter ($\frac{1}{12500}$ in.) and silt as any material the particles of which are above this but below ·04 mm. in diameter ($\frac{1}{625}$ in.): in the

United States, however, the limits are ·005 mm. ($\frac{1}{5000}$ in.) and ·05 mm. ($\frac{1}{500}$ in.) respectively.

A second group of soil constituents includes those derived from the organisms deposited along with the mineral particles while they lay under water. Of these the most important is calcium carbonate, the substance of which chalk and limestone are composed, and into which lime rapidly changes when added to the soil. As we shall see later on it plays an extremely important part in the soil, and profoundly influences the fertility and the vegetation relationships. It differs from the substances already described in that it dissolves somewhat in the soil water, and notable quantities are washed out, amounting at Rothamsted to about 800 lbs. per acre per annum.

Calcium phosphate also belongs to this group, although some is derived from rock. It commonly occurs only to a small extent, but it is an indispensable food for plants and therefore essential to fertility.

These two groups of inorganic substances—the silica and complex silicates derived from the rocks, and the calcium carbonate and phosphate derived in part from organisms that once have lived—do not form the complete soil. A third constituent is present, the so-called organic matter, derived from previous generations of plants. It is a familiar observation that no ordinary soil remains long without a covering of some sort of vegetation. As this dies its residues mingle with the mineral particles, being carried in by earthworms and various insects. The effect of this addition is very great. In the first place it profoundly influences the amount of plant food in the soil. The first vegetation that sprang up must obviously have got its food—its calcium and potassium salts, phosphates, etc.—from the mineral particles, but new sources of food appear for the plants that come after. The first crop slowly decayed under influences we shall deal with later on, and in decaying it set free those substances that its roots had taken as food and returned them again to the soil. Hence subsequent plants have food from two sources : the potassium salts, etc. dissolved by the soil water from the soil particles ; and in addition a supply of the same substances drawn by previous generations from the soil during their lifetime, but afterwards set free on the decay of the dead tissues. The plant food, in fact, keeps circulating between the soil and the plant, and the organic matter constitutes the medium by which the circulation is completed.

The second effect of the organic matter is even more important. During its lifetime the plant has been making a good deal of the substance of its leaves and stems from the gases of the air and the rain water, and the materials thus formed contain stored up energy derived from the sunlight. When they mingle with the soil and begin to decay the energy is liberated in the form of heat, and by the time they are completely decayed they have given out just as much heat as if they had been burned in a bonfire. The original heap of mineral matter contained no easily available store of energy ; the mixture of mineral matter and plant residues on the other hand does. The consequence of this addition is very profound : life is now possible in the soil, and there springs up a vast population of living creatures all drawing on this accumulated store of energy, flourishing so long as it holds out, and dying off when it is exhausted.

In our climate, and in humid climates generally, the decay of the plant residues is not complete, at any rate during the course of a few seasons, and some of the products accumulate as dark brown or black substances conveniently known by one name, humus. These substances have certain physical properties which they impart to the soil ; they are sticky, they absorb and retain water, they swell when wet and shrink when dry. In other words they are colloids. Thus the third effect of organic matter on the soil is to increase the amount of colloidal material, but some of this is of entirely different character from that already present. By far the most significant of these effects, however, is the bringing in of stores of energy because this constitutes the vital distinction between a heap of mineral matter and a soil. There is no soil without life, and no life is possible without stored up energy. We are only beginning to know what this soil life is, but already some hundreds of different kinds of creature have been found. Some few are large enough to be seen. Of these the most important are the earthworms, which burrow in the soil and effect a fine natural cultivation, letting in air and drawing in leaves, stems, and other vegetable debris from the surface to mingle with the mass of soil below. Most of the soil organisms are microscopic in size ; some lead an active life, others are in the inert resting stage and are called spores or cysts. The very incomplete census taken so far shows that the numbers of micro-organisms living in a single salt-spoon full of soil must be reckoned in millions.

Some of these organisms—certain bacteria—play a controlling part in soil fertility because they bring about the decay of the plant residues and consequent liberation of plant food. Out of the old dead plants, in fact, they make food for new ones. Thus the new generation of plants is dependent on them, just as they in turn are dependent on the past generations of plants. As more and more knowledge is gained the circle of soil life widens out and other varieties of organisms are seen to come in, interacting one on the other, not all making plant food, but all dependent in the last instance on the energy stored up in the organic matter, in other words, on sunshine that was caught years ago by plants long since dead.

There is reason to suppose that the four great constituents of the soil—the inert fragments of sand, the reactive clay, the calcium carbonate and the organic matter—are not merely lying alongside of one another in the soil. The evidence indicates that the colloidal constituents form a jelly-like coating over the inert particles, and this jelly contains much of the food of plants and of bacteria ; it may be likened to the nutrient jelly of the bacteriologist.

The soil mass is not solid throughout but is full of pores like a sponge. In a compact arable soil not more than 60—70 °/₀ of the volume is soil material, the remaining 30 or 40 °/₀ being empty space ; in a pasture soil the proportion of empty space is even greater. This pore space is at times completely filled with water, but more usually air is also present : at Rothamsted often to the extent of 10 °/₀ of the volume, leaving 25 °/₀ filled with water. The water is not pure but contains in solution carbonic acid, nitrates, carbonates and other salts of calcium, magnesium, etc.; it is held in the soil partly by surface attractions and partly by the colloids.

We may thus think of the soil as a porous mass made up of a hard framework plastered over with a jelly containing chemically active substances, plant foods, and unstable organic compounds rich in stores of easily liberated energy, while the pores contain air and a considerable amount of water.

Into the pores of this mass we have no means of penetrating : no microscope has been devised that enables us to look into it and see what is going on. We have indirect but incontrovertible evidence, however, that it is full of life and that the soil is inhabited by myriads of organisms of very varied kind, some of which, like eel-worms, are easily visible with a small microscope, while others, like bacteria, require a high power to reveal their presence. They bring about decay, and thus clear away the residues of previous plants leaving the soil clear for a new race. They do even more : they make the old plant material into new plant food. There are signs of conflicting and competing groups of organisms, but all at any rate have this in common : that they are dependent absolutely and entirely on the organic matter of the soil.

In its main outlines this conception of the soil is probably correct, and every month adds to our knowledge of the details. But the picture is still far from complete and it does not enable us to explain all the facts about the soil that have been gleaned by good farmers and gardeners. Each important discovery that is made opens out a wider field for exploration, and we may be certain that we never shall know all about the soil.

We shall find the study of the soil very unsatisfying and uninspiring if we become too much absorbed in its utilitarian aspects and forget to stop and reflect on the infinite wonder of its honeycombed structure and its dark recesses, inhabited by a teeming population so near to us and yet so hopelessly beyond our ken that we can only form the dimmest picture of what the inhabitants are like and how they live.

CHAPTER II

HOW PLANT FOOD IS MADE IN THE SOIL

By far the greater part of the food of the plant comes from the atmosphere : oxygen, carbon dioxide and water between them furnish most of the material out of which the plant is built. But it was discovered long ago that something is taken from the soil, and that this part, although small, is absolutely indispensable to the growth of the plant. The food thus furnished by the soil is really composed of a number of substances, the most important of which are nitrates, phosphates and other salts of potassium, calcium, magnesium, sodium, etc. It is convenient to divide these into two groups, the nitrogenous group, such as the nitrates, and the mineral group, including the phosphates, etc. of potassium and other metals. There is good ground for the distinction. The nitrates are derived almost exclusively from organic matter, but the mineral food, on the other hand, comes partly from the rock material of the soil. Further, the nitrates are easily soluble in water, and, therefore, readily washed away; they are, besides, liable to other sources of loss, while the mineral food only suffers slight losses. Lastly—and this aspect cannot be overlooked in a technical subject like ours— nitrates and other nitrogenous foods are by far the most expensive when any purchasing has to be done.

In trying to find out how plant food is made in the soil, investigators have confined themselves almost exclusively to the nitrogenous portion. This restriction was forced on the earlier workers by the circumstance that our soils stand much in need of nitrogenous manure: very much of the Rothamsted work has been and still is devoted to the study of the nitrogen problem, and a large part of our present knowledge is built up on the foundations laid by Lawes, Gilbert and Warington. We also must be content to accept the restriction, and pass over any changes which the mineral food may undergo, for the very good reason that we know so little about them.

It was early discovered that the plant residues

or farm-yard manure (which is essentially the same thing) added to the soil are not the actual food of plants, but only the raw materials out of which food is made. The true food is the nitrate to which the organic matter gives rise, and our first business is with this.

Owing to the losses which the nitrate suffers there is rarely any great stock of it in the soil, frequently not enough for the current season's growth. Fortunately the process of nitrate production, commonly called nitrification, goes on fairly readily so that fresh supplies are forthcoming whenever the conditions are suitable. The process has turned out to be very wonderful. It was formerly supposed to be entirely chemical, but a remarkable piece of work by Schloesing and Müntz in 1877 showed that it was brought about by bacteria. In studying the purification of sewage by land filters they caused a stream of sewage to trickle slowly down a column of sand and limestone, the experiment being continued for some weeks. For the first 20 days the ammonia in the sewage remained unaltered, then it began to change into nitrate, and finally the issuing liquid contained no ammonia but only nitrate. Why, asked the authors, was there this delay of 20 days before nitrification began? If the process were a purely chemical oxidation it should begin at once. If, however, it were bacterial, they could readily explain the delay, because the organisms would have to grow. To test this hypothesis they added a little chloroform vapour and found that nitrification was stopped entirely: it could, however, be started again by adding a little turbid extract of fresh soil after the chloroform was removed. They concluded, therefore, that nitrification was the work of "organised ferments." More rigid proof was afforded by Warington and later on by Winogradsky.

More recent experiments render it highly improbable that any chemical or physical process going on in the soil gives rise to nitrates, and we may take it that their production is entirely bacterial. Warington showed that the process takes place in two stages ; the ammonia is first converted into nitrites by one organism, and the nitrite is then changed to nitrates by another organism. Nobody has yet succeeded in finding any third stage between ammonia and nitrites although one might be expected on chemical grounds. There is practically no waste of ammonia during the process, and the conversion is almost if not entirely complete, but its mechanism is not at all understood and it cannot be reproduced artificially.

The organisms which alone can bring it about are utterly unlike any others and completely baffled the earlier investigators. Bacteriologists usually grow their organisms on gelatine or some similar medium. But this plan invariably failed to bring out the nitrifying bacteria, and it was not till Winogradsky in 1891 hit on the brilliant idea of using a jelly of silica, that they were grown and studied. Both organisms are extremely small—they are, in fact, the smallest known in the soil. Unlike the others they do not require organic matter as food, they make their own supply from carbonates or, like plants, from carbon dioxide. But unlike plants they do not want sunlight for this purpose, indeed sunlight kills them. Where, then, do they get their energy from? Winogradsky adduced very strong evidence, which has never been disproved, that the energy comes from oxidation of ammonia : he found a definite relationship between the amount of ammonia oxidised and the amount of carbon assimilated. It appears that ammonia is the only compound they can utilise. Many other substances have been tried, but without results, and we can take it as proved, as well as any negative proposition can be proved, that ammonia is the only substance from which nitrates are made, and that all the nitrate we find in the soil has previously been ammonia. This conclusion is very important and leads us to look for ammonia in the soil. But in no arable soil yet examined has more than a trace been discovered at any time of the year. We must, therefore, conclude that the rate at which ammonia is oxidised to nitrite is greater than the rate at which it is formed. But nitrites are never found in normal soils. It therefore follows that the rate at which nitrites are oxidised is greater than the rate at which they are formed.

We have, then, three reactions going on:

Nitrogenous plant residues change to ammonia,
Ammonia changes to nitrite,
Nitrite changes to nitrate.

Of these the last is the quickest, the second is slower and the first is slowest. The first change, therefore, limits the rate at which nitrates are produced : if we could speed up this change we should hasten the others.

Further, the quickest change (the third) is most susceptible to external influences. The organisms are very sensitive, they are more easily killed than the rest and they stop working more readily. Ammonia producers, on the other hand, are very resistant and will tolerate somewhat rough treatment. All three compounds, ammonia, nitrites and nitrates can be used as plant food, but in normal conditions the plant does not get the chance of using anything but nitrates, the other two being only transitory products.

Four important facts have thus been established with regard to nitrification :

(1) Ammonia is changed to nitrite and this is then changed into nitrate in the soil, the conversion being almost complete.

(2) Nitrates are formed from ammonia alone, and not from any other substance so far as is known.

(3) Nitrate production is the quickest of all the chain of processes.

(4) The quickest acting organisms seem to be the most sensitive.

We now turn to the formation of ammonia. This has not proved so attractive a subject for investigation and we do not yet know much about it. Unlike nitrification it is not a specific property of any one organism, but is effected by many different kinds: it is not even confined to bacteria, but goes on in a vacuum or in presence of antiseptics. But it will not go on after the soil has been heated to 150°, whence we may conclude that one of those accelerating agents technically known as enzymes is also at work. It is difficult to say precisely how much of the decomposition is due to enzymes free in the soil and how much to micro-organisms, but it seems certain that the latter are by far the most potent agents. No one has yet gone further back in the chain to discover the compound antecedent to ammonia. The initial compound is the nitrogenous part of the plant residues or of the substances added as manures, and is generally of a protein nature: enzymes, earthworms, fungi and bacteria may all take part in its decomposition ; we may suppose that its conversion into ammonia will go on in substantially the same way in the soil as it would elsewhere: but there is no definite evidence in proof.

The nitrogen compounds of the soil, however, are not entirely changed into nitrates: a second action takes place that is not in the least degree understood. When protein and other nitrogenous plant compounds are decomposed by micro-organisms in presence of air there frequently appears to be a large amount of gaseous nitrogen given off, especially when much organic matter is present. This change has been very little studied and indeed is commonly confused with another that appears to be wholly distinct— the so-called denitrification, a reduction of nitrates brought about by certain bacteria in presence of organic matter, but in *absence of air.* Whatever its nature it leads to great losses in rich or heavily manured soils, and is responsible for much of the exhaustion of virgin soils that is now going on at an appalling rate. One of the most pressing problems before the agricultural chemist is to study these two sets of reactions, and in particular to find out whether this wasteful process cannot be suppressed, so that a larger part of the nitrogen compounds shall change into the useful nitrates. In modern farming nitrogenous manures are by far the most expensive, and profits are cut so low that all sources of loss are to be avoided as far as possible. Both these actions, the formation of nitrates as well as the loss of nitrogen, tend to use up the nitrogen compounds of the soil, because the nitrates not taken by the plant are speedily washed away and lost in the rivers and the sea. As the original stock was probably never high, it is clear that there must be some reverse process by which the soil gains nitrogen, or the supply would long since have given out. This has long been realised by men of science, and a careful and systematic search begun 30 years ago was ultimately rewarded by the discovery of two ways in which such a gain takes place.

The old-established cultivator of the land has a great stock of information about the ways of plants ; some of it is disconnected and fragmentary, but it has to be sorted over and examined experimentally by the man of science. One of the old bits of knowledge handed down from time immemorial, and already traditional when Virgil wrote his *Georgics*, was that beans, vetches, and lupins improve the land for the next crop. Sow your golden corn, says Virgil, on land where grew the bean, the slender vetch or the fragile stalks of the bitter lupin[1]. When Lawes and Gilbert began their experiments in 1843 one of the early problems was to discover the reason for this improvement, and they were able to trace it to the fact that a soil was richer in nitrogen after the

---

[1] *Georgics*, Book I, lines 73 *et seq*.

growth of clover than before. Somehow or other
the amount of nitrogenous food in the soil had increased. But no one could give any satisfactory
explanation *why* the nitrogen should increase, and it
was not until 1886 that the solution was found. The
story is so interesting that it must be told again,
although it has often been told before.

Hellriegel and Wilfarth, two distinguished investigators at the Experiment Station at Dahme, in
Prussia, were studying the effect of nitrates on plant
growth and found that the amount of growth of
cereals like barley, oats, etc., increased as the nitrate
supply increased and was, in fact, directly propor-
tional to the amount of nitrate. In the case of lupins
and allied plants, however, no sort of proportionality
could be traced, the plants sometimes did as well
or better without nitrate as with it, but sometimes
failed altogether. Further, chemical analysis showed
that the quantity of nitrogen present in the cereal
crops was just about the same as that supplied,
while the quantity present in those peas which made
any growth was much greater. It followed, therefore, that these peas had got some of their nitrogen
from the air. But why had not all the peas done so?
Hellriegel and Wilfarth argued that the success of
the process must depend on something that only
came into the experiment by chance. At that time
men had bacteria very much in their minds because

of certain wonderful discoveries that had recently been made. Hellriegel and Wilfarth, therefore, very naturally asked if bacteria could be the active agents here, particularly as they knew that the little swellings on the roots of the pea—the so-called nodules— contained bacteria, and also that some bacteria could take in gaseous nitrogen and use it. To test the matter peas were sown in sterilised sand (i.e. sand baked so as to kill all living organisms), containing mineral food, but no nitrogenous food; these made little or no growth and developed no nodules in the roots. Other peas were also sown in similar sand, but they received a water extract of ordinary arable soil ; these made excellent growth and had a marked development of root nodules. If, however, the extract was first boiled it had no effect in increasing growth.

These experiments afforded satisfactory evidence that the pea could form an association with certain bacteria which should be self-supporting so far as nitrogen was concerned in that it could draw on the immense stores of free nitrogen in the air. The proof was made more rigorous by other and later workers, and the proposition is now one of the most definitely established in modern science.

Thus peas, vetches, lupins, and, we can add, beans, clover, lucerne, sainfoin, in short all the tribe of the leguminosae, take in stores of nitrogen from the atmosphere through their association with the bacteria of the root nodules. When their roots, leaves or stems perish and mingle with the soil these newly-furnished nitrogen compounds are added to the general stock already there.

A few years later another set of bacteria was found able to take in and use the free nitrogen of the air, differing from the preceding in that they work on their own account and do not form associations with plants. The earliest to be discovered was Clostridium, but interest has centred largely round a later find, Azotobacter, because it works in presence of air and not, as happens with certain other organisms, in its absence.

The more one studies these nitrogen-fixing organisms the more remarkable do they appear. The absorption of gaseous oxygen by living organisms and the changes it brings about can be paralleled more or less closely by artificial processes in the laboratory. But the absorption of gaseous nitrogen by these particular organisms cannot be imitated in the laboratory and is without parallel in our experience. A source of energy is needed, and a considerable number of substances are known to serve, including sugar, starch, cellulose, or residues of plants. As the organisms are very widely distributed they may be expected to operate wherever supplies of easily decomposable organic matter are present in the soil,

e.g. wherever vegetation is allowed to die back. It
has been found by actual measurement at Rothamsted
that nitrogen does accumulate in soil left to run wild
and to cover itself with the varied assortment of
plants cropping up in these conditions. How much
of this is due to Azotobacter is not certain, because
leguminous plants occur among the herbage and fix
an unknown quantity of nitrogen.

The organisms that we have been considering
represent the constructive agencies in the nitrogen
cycle in the soil, bringing in new supplies from the
air and so making good the losses already discussed.
It is necessary to remember—this point will constantly recur in future chapters—that *these constructive processes only manifest themselves in soils
covered with permanent vegetation* such as grassland, woodland, etc. One exception only is known,
viz. where a leguminous crop is growing, when the
amount of nitrogen fixed may be considerable. With
this exception one general rule holds : losses of
nitrogen preponderate on soil that is cultivated, and
gains of nitrogen preponderate on soils covered with
permanent vegetation. In either case the action
does not go on indefinitely: the losses become less
and less as the soil becomes poorer, till finally they
are so small that it is difficult to detect them, and
the gains also become less and less as the soil becomes richer, till finally they also cease or are

balanced by losses. Thus limits are finally reached,
and no soil becomes absolutely destitute of nitrogen
or very rich in it; few, if any, of our British soils
(leaving out sand dunes and peat bogs, which are not
true soils) contain less than 0·05 per cent. or more
than 1·0 per cent. of nitrogen.

We may summarise the changes of the nitrogen
compounds of the soil in diagram form thus:

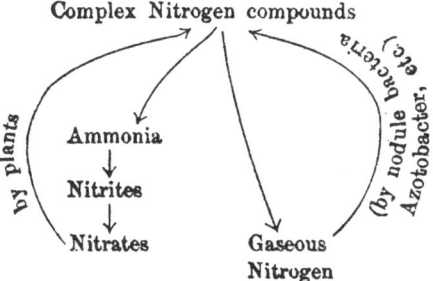

The conversion of the complex nitrogen compounds
into nitrates is the process whereby plant food is
made, and the fixation of gaseous nitrogen is the
means whereby the stock of nitrogen compounds is
maintained. Both these are obviously indispensable
to plant growth and to the fertility of the soil. The
other change, the evolution of gaseous nitrogen from
complex nitrogenous compounds, appears on our
present knowledge to be sheer waste and to serve

no useful purpose whatsoever. Whether or not further knowledge will show that it is really an essential part of the scheme we cannot say; our endeavour now is to reduce it as much as possible.

More recent investigations made at Rothamsted and elsewhere have brought out the striking fact that conditions which are injurious to active life in the soil often bring about *increases* in the numbers of bacteria and in productiveness, while conditions favourable to active life often lead to *decreases* in bacterial numbers and in productiveness.

This apparent paradox was solved by showing that two groups of organisms occur in the soil: the useful makers of plant food, and another set detrimental to them but, fortunately, more easily killed and slower in multiplying. When adverse conditions appear the detrimental forms suffer more than the useful forms : thus long severe frost, hot dry summer, heat, treatment with mild poisons that can subsequently be removed, all put them out of action temporarily, if not permanently, and so lead to greater bacterial activity and greater productiveness. The detrimental forms are provisionally identified with the protozoa in the soil, of which numbers have now been found.

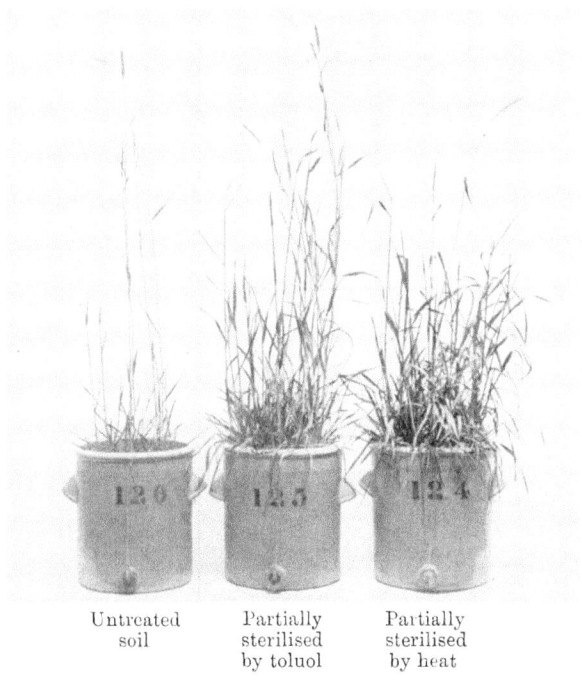

Untreated	Partially	Partially
soil	sterilised	sterilised
	by toluol	by heat

Partially sterilised	Untreated	Partially sterilised
by toluol	soil	by carbon
		disulphide

Fig. 1. Crops grown on untreated and on partially sterilised soil

CHAPTER III

WHAT IS SOIL FERTILITY AND HOW MAY IT BE ATTAINED?

THE relation between the soil and the plant is not entirely simple, and in consequence no rigid definition of soil fertility can be given. Any particular soil would probably prove very fertile for one at least of all the thousands of plants in existence, but if it were useless for ordinary agricultural and horticultural purposes it would generally be called barren. A fertile soil may be described as one in which the conditions are favourable for the growth of plants useful to man.

Six requirements are necessary for the plant: water, air, temperature, food, root room, and absence of harmful factors. We must now briefly discuss these in their relation to the soil.

Water supply. The soil receives water from the rain and from the subsoil, but it also loses water by evaporation and drainage. The actual amount present in the soil at any time therefore depends on several factors as it obviously represents the balance of gains over losses. The amount of rainfall is important, but its distribution is even more

so, in determining the amount reaching the soil
(heavy rains being liable to run off the surface while
lighter rains soak in), and the temperature and
wind are great factors in determining how much
stays there. February is one of the driest months
so far as actual rainfall goes, but no one who lives in
the country need be reminded how persistently wet
the ground generally is then. August, on the other
hand, is one of the wettest months, but the soil is
usually dry. Equally important factors are the nature
of the surface soil, which determines how much of
the rain percolates; the position of the soil in respect
of the surrounding land—whether high or low lying—
and the nature of the subsoil, which regulate the flow
of the underground water.

A defective water supply may therefore be either
the fault of the soil or its misfortune. Too much
clay or peaty organic matter may render the soil
impervious to rain and so cause it to lie waterlogged,
while too much coarse sand or chalk may keep it so
open that water rapidly drains away or evaporates;
in all these cases the soil is to blame. On the other
hand, the very best soils may remain stagnant marshes
if they occupy low-lying ground on which water
drains from the hills without finding an outlet; or
they may suffer badly from drought if they are
spread out thinly on beds of gravel or of rock. The
stagnant marsh may be drained and the soil soon

reveals its true nature; the rock, if thin enough, may be broken or removed; but the thinly spread soil on a gravel bed is beyond our present powers of treatment.

When the cultivator has any reason to doubt the efficiency of the water supply to his plants he must first ascertain whether the fault is in the soil or its surroundings. The bad effects of too much clay may be modified by liming, chalking, or drainage; excessive sand may be counteracted by additions of clay or silt or by ploughing in organic manures and green crops. In all cases the water supply is improved by extending the range over which the roots may grow, for the larger the volume of soil on which the plant draws the less will be the amount of water required from each portion. And so if any obstacle exists to the development of the roots it must be removed; the soil must be dug or ploughed more deeply and, in a garden, manure must be added to the lower spit; any pan or thin rock layer must be broken up and removed; and stagnant soil water must be prevented from rising too high by tapping the springs or by laying deep drains[1]. A good deal can be done by skilful cultivation to check evaporation and thus reduce the loss of water. A fine layer of soil on the surface effectually shields the rest of the soil from the sun's heat and keeps the moisture safe from loss.

[1] See p. 78.

The constant use of the hoe in the garden reduces considerably the need for watering, and many a good gardener will declare that "the hoe is the best watering can." In dry regions the disk-cultivators are brought out as soon as possible after a shower so as to break up any crust that may have formed on the arable land and restore the protective coating of fine soil.

Air supply. The harm that results from a water-logged soil is not due to the excess of water but to the exclusion of air. The plant roots and the food-making bacteria alike need air, and air must therefore be allowed access to all parts of the soil. Once the excess of water goes the air comes in: such devices as liming a clay soil, laying drains, and breaking a pan, therefore have the effect of improving not only the water supply but the air supply as well.

Temperature. The heat relationships of a soil are also intimately bound up with the water content. Its mean temperature is of course directly dependent on its location, and does not differ greatly from that of the atmosphere. The top half inch of soil is hotter than the air in direct sunshine and colder by night: a little below the surface the fluctuation is no greater than in the air, while six inches below it is much less. Heat only travels slowly through dry soil and does not affect the subsoil till some time after it has reached the surface. But it travels more quickly through moist soil; there is less lag in transmission and hence less accumulation of heat in the top layers. Dry soils are therefore hotter at the surface in sunny weather, but moist soils are the more uniform in temperature. April showers do much to warm the subsoil by setting up heat communication with the surface and enabling the warmth of the sunshine to travel down; on the other hand, winter rains cool the subsoil by letting out the heat stored up there. Another important factor is the difference in specific heats: water requires five times as much heat to raise its temperature through 1° as does dry soil, so that a given quantity of sunshine is less able to raise the temperature of a wet than of a dry soil.

Besides these factors concerned in the warming of the soil there is another that has the effect of cooling it. Heat is required for the evaporation of water, and in consequence the soil is cooled when any of its moisture dries off.

Several devices may be adopted in warming the soil. The heat received may be concentrated on certain parts of the soil by laying it up in ridges running E. and W. and therefore facing S. The amount retained (for some is always reflected back into space and lost) is increased by dressing with a layer of black material, soot being the most convenient. The effectiveness of the heat received is increased by draining away any excess of water.

Finally the losses may be reduced by checking evaporation, and this can be done by sheltering from the wind with windbreaks or hedges and by maintaining the fine tilth at the surface. Horticulturists sometimes adopt all these, agriculturalists sometimes only one or two.

Food supply. The supply of plant food in the soil depends in the first instance on its mineral composition. The great mass of inert material that constitutes the framework of the soil and subsoil affords but little food. The food constituents are to be sought among the more soluble and reactive substances, and, in some way that is not sufficiently understood, their availability is increased when calcium carbonate is present. A second factor, the accumulation of plant residues, began to come into play soon after the soil was formed and has in some cases assumed so much importance that it now controls the situation. And, as plant residues are not themselves plant food, but have to be converted by micro-organisms into simpler substances, we can add as a third factor the activity of the food-making organisms.

It is a simple matter to increase the supply of plant food by adding fertilisers to the soil. The amount of nitrogen may be increased by adding nitrate of soda, sulphate of ammonia, organic substances such as farmyard manure, guanos, certain manufacturers' waste products, etc. or by growing leguminous crops. Since the loss of nitrogen is considerable recourse must often be had to one of these methods, and grave doubts have at times been expressed as to how long the world's supply of nitrogen compounds would last. Of late years, however, it has been found practicable to make nitrogen compounds from the inexhaustible stock of nitrogen in the air; the world's supply of nitrogen manures can therefore be increased whenever necessary, and the dreaded nitrogen famine has been relegated to the time when the energy supply shall give out.

Phosphates are supplied in the form of bones, guanos, basic slag and rock phosphate. Most soils contain insufficient to satisfy the large crops produced by the modern farmer; frequent additions are therefore necessary, especially under conditions of high farming. The losses are not as serious as those of nitrogen. There is practically no washing out; the agricultural chemist can still detect the phosphates added to the lands round the cities of ancient Egypt. Nevertheless there is a steady and continuous loss in the crops which has to be made good. There is no way of adding a single ounce to the world's stock of phosphorus compounds, and this is being drawn upon even now to the extent of some millions of tons each year, while the demand increases steadily as time goes on. If some day the supply gives out, and at present it seems inevitable that it must, mankind will be faced with perhaps the most serious of all catastrophes, a phosphorus famine.

But there is no immediate cause for alarm, as on the lowest computation the visible supply will last for many years. Further, the phosphates lost from the farm are not destroyed but find their way to the sewage, and thence to the sea. We must therefore look to the ocean for the means of replacing the land deposits of phosphates, and already a fair amount is drawn from this source in guano, fish meal, etc.

The supplies of potassium compounds already existing in the soil are in general sufficient for ordinary purposes, but additional supplies become necessary as the character of the farming improves and larger crops are grown. Some special soils, such as peats, chalks, and thin sands, need potassium fertilisers in order to yield even small crops, and much labour has been lost through ignorance of this fact. The deposits of potassium salts are extensive and in Germany's keeping; no needless waste is therefore to be apprehended. In the last instance the ocean can be made to give up some of its enormous stock.

The nature of the plant residues and the ease with which they are decomposed by bacteria depend on climatic factors—the temperature, water supply, etc.—and also on the amount of calcium carbonate present. These various relationships have already been discussed at some length: we need now only point out that a water supply suitable for ordinary plant growth seems also to be very suitable for bacterial activity. The rule seems to apply also to other soil conditions, and we may make the general statement that a soil suited to the growth of plants is also suited to the activity of bacteria and therefore to the production of plant food. The similarity becomes even more close after the soil has been partially sterilised so as to destroy detrimental organisms.

A further connection between the food supply and the water supply lies in the fact that the food has first to be dissolved in the soil water before it can enter the plant. Roots have no power of taking in solid matter; they can only imbibe solutions; further, they cannot use strong solutions, but may even be injured thereby as in some of the alkali lands; and they do not make very good growth in too weak solutions.

Even this is not all. The amount of plant food per unit volume of the soil is not the only factor determining the amount of food the plant can get; the extent of the root range is equally important just as it was in the case of water.

Root range. And this leads to the question of root room. No plant does well unless it has ample space for the full development of its roots. It suffers from the restriction of its supply of water and of food, and apparently from other causes as well; recent experiments seem to indicate that new factors may come into play when one set of roots runs up against others because there is not space for both. In Mr Pickering's experiments at Woburn growing grass had a very detrimental effect on the fruit trees planted in it, and there is also evidence that weeds may have a directly harmful action on sown crops.

Absence of injurious factors. However good its food and water supply, a soil may remain infertile if injurious substances happen to be present. Little attention has been paid to these in England, but they have been much studied in the United States where, indeed, they have given rise to considerable controversy. It seems beyond dispute that substances harmful to plants do occur in wet soils poor in calcium carbonate. Nothing is known about these substances (in English soils, at any rate) but some years ago the fashion arose of calling them acids without any sufficiently rigid proof. It is distinctly unwise to prejudice future investigations by assigning a name already used for a definite group of substances to another that is yet unstudied, and we shall therefore adopt the nomenclature of the practical men and speak of such soils as "sour." Whatever the cause of "sourness" (and the name commits us to no hypothesis), it can be remedied by drainage, lime, and good cultivation.

Cases are recorded of infertility arising from excess of iron or of manganese in the soil, but no satisfactory evidence is afforded in proof. Drainage, lime and good cultivation are here also found to be beneficial.

"Sickness" of soil has in the past been attributed to the presence of a toxin, but more recent work indicates that it is biological in nature and remedied by partial sterilisation.

The alkali soils of dry regions owe their sterility to excess of soluble salts; they may be treated by drainage (which is a first essential), irrigation, and addition of gypsum where much sodium and potassium carbonates are present. Much interesting work remains to be done in elucidating the causes of infertility of certain special soils.

Looking back over these various fertility factors we see that they are not a mere tangle of unrelated things, but are very closely connected one with the other. The water supply, air supply, and temperature are to a large extent mutually interdependent, and changes in any of them are reflected in the food supply. We can simplify matters by selecting three as the leading factors in normal cases: the water supply, food supply, and stock of calcium carbonate; when these are satisfactory it will usually be found

that the other three conditions are also favourable
to plant growth.

Soil Types.

In order to increase the soil fertility it is necessary
first to seek out the factor limiting plant growth and
then to remove it. As the different soil types have
certain characteristic limiting factors we can now
advantageously turn to them for a time.

Sandy soils consist chiefly of inert silica with only
about 6 per cent. or even less of clay, and are con-
stitutionally poor in those mineral compounds that
give rise to plant food. Any stock that might
originally have been present is constantly being
reduced by solution in the rain water that drains
through. They are therefore poor in plant food.
But on the other hand, if they are free from such
obstructions as layers of rock, hardpan, or stagnant
water, they allow of very copious and deep development of plant roots. The actual volume of soil upon
which the plant may draw for food and water is considerable, and in consequence sandy soils yield better
crops than might at first be supposed. Generally
also the crops ripen well and give early produce of
good quality. Only small amounts of calcium carbonate appear to be necessary to prevent sourness.

The deficiency in food is readily made good
by frequent small additions of manure. The water

supply tends to be erratic because of the great ease
with which rain water soaks into the depths of the
soils, but it can be made more regular by additions
of organic manures or of clay. The defects of a sandy
soil are mainly negative, i.e. they can be remedied
by adding something, whilst its advantages are very
real; it induces good root development, early yield
and high quality. There is, perhaps, a wider range
of possibilities for a sandy soil than for any other;
it may be a desolate heath, or it may, under proper
management, blossom out as a fruit and vegetable
garden, giving each year two or three crops of good
produce. But the cost of the process may be more
than the result is worth.

Soils that contain more of the fine clay material
and proportionally less sand are called loams. It is
impossible to define loams exactly: the cultivator
recognises them by the fact that they are definitely
coherent and not loose like sand, yet not over-sticky;
and further that they allow free root development.
But there are no sharp lines of demarcation; many
soils at one end of the scale would be called light
loams by some practical men and sands by others,
while at the other end soils called heavy loams by
some would be regarded as clays by others. In
between these limits there remains a great body of
soils which most cultivators would agree to call
loams, and these on analysis are found to contain

6 to 15 per cent. of clay, 40 to 60 per cent. of the silts and 20 to 50 per cent. of coarse and fine sand.

Loams contain more plant food than sands and in general have a better water supply; they therefore yield heavier crops. But the crops are often not as early as those grown on sands.

The next group, the clays, cannot be sharply marked off from the loams but can only be described generally as sticky soils, persistently wet in winter and spring, and drying to hard clods in dry weather. They contain more actual clay[1] and less sand than the loams, but none appears to contain more than 50 per cent. of clay, very few contain as much as 40 per cent., and most "clays" contain only about 25 per cent. or less. In consequence of their stickiness clays do not allow very free root development. The root range being thus restricted, plants do not draw on anything like the same volume of soil for food and water as in the case of loams and sands. Hence the clays are less fertile than these soils in spite of the fact that they often actually contain more food and water. Of course if the plant is in

[1] Unfortunately soil chemists use the word "clay" in two distinct senses: (1) the soil or mineral as a whole, (2) the fine material less than ·002 mm. in diameter (·005 mm. in the United States). In past years another meaning was given which does not appear to have been very definite; this survives in the statement handed solemnly down through eighty years of text books that a clay soil contains "75 to 95 per cent. of clay." Ceramic chemists adopt a different definition.

the ground a long time it can develop a big root
system and then it will grow well: trees, grass and
the longer-lived deep rooting arable crops, particularly wheat and beans, do very well. Each of these
types of soil has its advantages: sands are easily
workable and present great possibilities in the way
of cropping: loams give good heavy crops of almost
any of the ordinary farm and market garden plants:
clays are very well suited to grass, wheat, and beans,
three crops of considerable importance to the farmer.
A really strict comparison is not possible because the
types are so different, but in the main we must give
the palm for fertility to the loams. The districts
in our own country famous for their fertility are
commonly loams; they are often alluvial deposits
bordering the sea, as in the Chichester district, or
lying in broad valleys, as in the vale of Evesham;
they are assured of an ample water supply by
their position, and favourably situated in regard to
climate.

But we must never forget that every soil will bear
some plants well although they may not happen to be
saleable at the time. Gervase Markham's list drawn
up in 1620 needs but little change to-day. "Ground
which, though it bear not any extraordinary abund-
ance of grass, yet will load itself with strong and lusty
weeds, as Hemlocks, Docks, Mallows, Nettles, Ketlock
and such like, is undoubtedly a most rich and fruitful

ground for any grain whatsoever. And also, that ground which beareth Reeds, Rushes, Clover, Daisies and such like, is ever fruitful in Grass and Herbage.... When you see the ground covered with Heath, Ling, Broom, Bracken, Gorse or such like, they be most apparent signs of infinite great barrenness....And of these infertile places, you shall understand, that it is the clay ground, which for the most part brings forth the Moss, the Broom, the Gorse, and such like; the sand, which bringeth forth Brakes, Ling, Heath, and the mixt earth, which utters Whinnes, Bryars, and a world of such like unnatural and bastardly issues."

It was largely owing to the circumstance that mankind was unwilling to pay for "Whinnes, Bryars, and a world of such like unnatural and bastardly issues" that the "mixt earth" was called infertile. Tastes alter—who would now accept Markham's description of heather as "only a vile filthy black brown weed"?—and a plant despised by one generation may be prized by the next. Perhaps the most permanent piece of advice one can give to the cultivator of a piece of poor land is to find out what plants it will grow well, decide which will be most profitable, then ascertain what is preventing the soil from growing these better, and, finally, if possible, remove the hindrance, whatever it may be. The gap between the ideal and the actual conditions for the crop may be narrowed from both ends: the plant may be altered somewhat by the breeder, and the soil conditions may be changed in some of the ways to be described later on. Nothing but disappointment, however, is likely to follow attempts at growing plants or crops unsuited to the soil conditions.

CHAPTER IV

SOIL FERTILITY AND SYSTEMS OF HUSBANDRY

WE have seen that there is a close relationship between the composition of the soil and the vegetation growing on it; an even closer connection can be traced between the fertility and the system of husbandry.

Virgin land covered with its native vegetation appears to alter very little and very slowly in composition. Plants spring up, assimilate the soil nitrates, phosphates, potassium salts, etc., and make considerable quantities of nitrogenous and other organic compounds: then they die and all this material is added to the soil. Nitrogen-fixing bacteria also add to the stores of nitrogen compounds. But, on the other hand there are losses: some of the added substances are dissipated as gas by the decomposition-bacteria, others are washed away in the drainage water. These losses are small in poor soils, but they become greater in rich soils, and they set a limit beyond which accumulation of material cannot go. Thus a virgin soil does not become indefinitely rich in nitrogenous and other organic compounds, but reaches an equilibrium level where the annual gains are offset by the annual losses so that no net change results. This equilibrium level depends on the composition of the soil, its position, the climate, etc., and it undergoes a change if any of these factors alter. But for practical purposes it may be regarded as fairly stationary.

When, however, the virgin land is broken up by the plough and brought into cultivation the native vegetation and the crop are alike removed, and therefore the sources of gain are considerably reduced. The losses, on the other hand, are much intensified. Rain water more readily penetrates, carrying dissolved substances with it: biochemical decompositions also proceed. In consequence the soil becomes poorer. But the impoverishment does not go on indefinitely: the rate of loss diminishes as the soil becomes poorer, and finally it is reduced to the same level as the rate of gain of nitrogenous organic matter. A new and lower equilibrium level is now reached about which the composition of the soil remains fairly constant; this is determined by the same factors as the first, i.e. the composition of the soil, climate, etc.

Thus each soil may vary in composition and therefore in fertility between two limits: a higher limit if it is kept permanently covered with vegetation such as grass, and a lower limit if it is kept permanently under the plough. These limits are set by the nature of the soil and the climate, but the cultivator can attain any level he likes between them simply by changing his mode of husbandry. The lower equilibrium level is spoken of as the inherent fertility of the soil because it represents the part of the fertility due to the soil and its surroundings, whilst the level actually reached in any particular case is called its condition or "heart," the land being in "good heart" or "bad heart," according as the cultivator has pushed the actual level up or not: this part of the fertility is due to the cultivator's efforts.

The difference between the higher and lower fertility level is not wholly a question of percentages of nitrogen, carbon, etc. At its highest level the soil possesses a good physical texture owing to the flocculation of the clay and the arrangement of the particles: it can readily be got into the fine tilth needed for a seed bed. But when it has run down the texture becomes very unsatisfactory. Much calcium carbonate is also lost during the process: and when this constituent falls too low the soil becomes "sour" and unsuited to certain crops.

The simplest system of husbandry is that of

continuous wheat cultivation, practised under modern conditions in new countries. When the virgin land is first broken up its fertility level is high; so long as it remains under cultivation this level can no longer be maintained, but rapidly runs down. During the degradation process considerable quantities of plant food become available and a succession of crops can be raised without any application of manure. In fact there is commonly more plant food than the crop needs and addition of manure gives no crop increase. Hence arises the idea that the land needs no manure, and the pioneer, fully occupied with the pressing needs of the moment, not only supplies none, but does not even put back into the soil any part of what he takes away. The grain is sold and the straw is either burnt or dumped into gullies. After a time the unstable period is over and the new equilibrium level is reached at which the soil will stop if the old husbandry continues. In this final state the soil is often not fertile enough to allow of the profitable raising of crops; it is now starving for want of those very nutrients that were so prodigally dissipated in the first days of its cultivation, and the cultivator starves with it or moves on.

> "O man, that from thy fair and shining youth
> Age might but take the things youth needed not."

Such is the history of many of the derelict farms in parts of the United States and such must inevitably

Fig. 2*a*. Crop grown without phosphates on a worn out
soil in Illinois

Fig. 2*b*. Showing what phosphates will do on worn out
soils in Illinois

be the history of many farms elsewhere so long as continuous wheat culture is adopted. It is futile to speak of land as inexhaustible: fertility is no more inexhaustible than any other form of capital. The pitiful thing is that so much of the loss is sheer waste: about one third of the plant food goes into the crop, the rest is lost beyond hope of recall as gas into the atmosphere or as saline matter in the drainage water and the streams. It is not the crop that exhausts the land but the continuous cultivation.

Fortunately recovery is by no means impossible, though it may be prolonged. It is only necessary to leave the soil covered with vegetation for a period of years when it will once more regain much of the nitrogenous organic matter it has lost. But it does not wholly recover. The phosphates and potassium salts removed in the crops, and the calcium carbonate leached out, are not regained; for want of them the growth of recuperative vegetation may suffer.

The problem has been investigated with characteristic energy in the United States, and a remedial scheme has been evolved by Dr Cyril Hopkins, Director Thorne and others, based on experience in the older countries and on careful experiments in the new. The central feature is that continuous tillage must stop, and for one third to one half of its time the land must lie untilled and covered with vegetation, i.e. in the course of six years not more

than three or four grain crops should be taken;
during the remaining time the land grows grass or
leguminous crops, so that it may gain organic matter.
It is not unremunerative during this period; on the
contrary, these crops are distinctly valuable. Much
the best for the purpose are clover, either alone or
mixed with timothy, and lucerne; these collect gaseous nitrogen as well as carbon dioxide to form
nitrogenous organic matter in the soil. It is an
indispensable part of the method that limestone
should previously be added to the soil, or the clover
or lucerne may fail. When the soil has been thus far
improved the supply of phosphates present may be
insufficient for the crop that can be produced, and
this limiting factor has therefore to be removed by
the addition of rock phosphates. The crop now
increases till it is limited by some new factor. It
may happen that the supply of potassium salts in
the soil constitutes this new limiting factor, in which
case addition of potassic fertilisers becomes desirable.
The fertility may then rise higher than it was at first
in the virgin soil.

The whole process, it will be observed, consists in
the successive removal of the limiting factors.

The exhaustion of the virgin lands constitutes the
simplest case because everything is removed from the
soil and nothing is put back. It could only arise under
conditions of cheap transport facilities between the

virgin land and the cities where the grain is to be consumed. Thus it is essentially a modern phenomenon; it never arose in so acute a form in England because such facilities did not exist till a complex system of agriculture was already established. Wheat has always been grown in this country and records exist of very early exports. Zosimus relates (*Hist. Nova*, lib. 3, ch. 5) that Julian brought wheat from Britain to feed the inhabitants of some of the Rhine cities whose stores had been destroyed and their harvest ruined by insurgent tribes. But we have no knowledge what the agricultural methods were. When the first definite records of English agriculture appear a system was already in use that kept the soil fertility at a sufficient level for the needs of the time.

In medieval England the arable land occurred partly in the lord's demesne and partly in the common cultivated field[1]. The latter was divided into fields, generally three in number, which were again divided into strips so distributed among the tenants that each should have his share of good and bad land. The pasture consisted of the common grazing land, certain outlying lands, and the cultivated common field after the harvest was off; in addition there were certain fields and water meadows not held in common. The live stock thus had a relatively wide area of ground over which to gather their food; and their

---

[1] Often called "infield" in the North.

manure, mingled with any bracken, straw, rushes, etc., gathered for litter, went to fertilise the arable land. The distribution was not very uniform, as the lord often had special claims, but all the arable land did receive some dung.

The manure having been put on, a crop of wheat or rye or both was taken. After harvest the individual cultivators no longer had any special rights in their strips and the whole field became common; the fences were removed and the cattle allowed to enter and graze the weeds and grasses. The ground was somewhat enriched by folding on it sheep that had grazed during part of the day on the common. This period lasted from Lammas Day[1] (August 1st) till Candlemas (Feb. 2nd). The land was next sown with barley, oats, or other spring or "Lent" corn, and after harvest (Lammastide) again grazed until the following Candlemas. It was then ploughed up and left fallow throughout the summer; finally it was dunged and sown with wheat. Occasionally, however, barley was taken first and wheat after:

> "First rie, and then barlie, the champion saies,
>   or wheat before barlie be champion waies;
>   But drink before bread corne with Middlesex men,
>   then lay on more compas, and fallow agen[2]."

[1] Lammas Day (Aug. 1st) may seem early to a modern farmer for the individual rights to cease and the whole field to become common, but it must be remembered that the Julian Calendar was then in force so that the date is really later than it looks.

[2] Tusser, *Five Hundred Pointes of good Husbandrie*, *Octobers*

Whatever the order, the rotation consisted of two corn crops and a fallow; each year one of the fields was fallow, the second wheat or rye, and the third Lent corn.

The dung made by the animals contained elements of fertility derived from the pasture land. The addition of this dung to the arable land thus involved a transfer of fertility from the wide areas of the pasture land to the smaller areas of arable land. The process maintained the fertility of the arable land, but it must in time have impoverished the pasture; but the impoverishment of the pasture went on only very slowly for two very interesting reasons. In England the supply of nitrogen compounds in the soil is most frequently the factor limiting the wheat and other grain crops ; so long as the nitrogen supply is kept up a certain level of crop production can be maintained. In pasture land a considerable amount of nitrogen fixation is continually going on through bacterial activity. Hence the element that played the most serious part in fertility under the conditions of low yields then obtaining was being brought in as quickly as necessary from the atmosphere.

The next substance to give out and cause the collapse of the system would have been calcium

husbandrie, 1573: drink corne = barley ; bread corne = wheat ; compas = farmyard manure (compost); champion = the unenclosed common field or its farmer.

carbonate. There was a real danger of this, but it
was met by chalking and marling which from time
immemorial had been part of the agricultural practice
of these islands (see p. 80). In consequence soil
exhaustion could only set in through exhaustion of
the phosphates and potassium salts, and this was a
very slow process which, moreover, was still further
delayed by the practice of fertilising with wood
ashes, which supplied potassium salts derived from
the forest, and less frequently with salt (sodium salts
economising the consumption of potassium salts by
the plant). Thus the only weak point in the system
was the exhaustion of phosphate from the soil, but as
the yield of wheat was probably often under 10 bushels
per acre, which would only take out some 5 lbs. of
phosphoric acid (P_2O_5), and as the total population
was only small and sparsely scattered, the system was
for all practical purposes permanent. But the yield
was very poor.

The next step up involved some very drastic
changes. Gradually the common arable fields and
pastures began to be enclosed and each man's holding
came into one piece. The process was slow and is
hardly complete even yet; the old strip farming of
common fields may still be seen in the Isle of Axholme,
Lincolnshire, while survivals of it may be detected in
many villages. For example, several arable common

fields can be traced around Harpenden; Manland, Westfield and Pickford Commons are divided by balks into strips as in medieval times. During the fifteenth and sixteenth centuries the enclosure was accompanied by the wholesale conversion of arable into grassland induced by the high price of wool, and both processes were much resented by the peasantry, who pulled down the new hedges in many places, and in Norfolk broke out into open rebellion under Kett in 1549. Shrewd writers of the period saw, however, that only on enclosed land could a higher level of productiveness be attained, and history has shown that only on such land were improved methods adopted.

Under the new conditions each man could grow what he liked (unless the landlord forbade him), and he was no longer tied down to follow ancient custom. There was a greater incentive to industry, more manure could be obtained and greater care could be taken with the cultivations and to keep down weeds. In consequence larger crops were now obtained; the yield of wheat has been estimated for certain districts at about 20 bushels, barley at 30, oats and pulse at 40 bushels per acre at this period. There can be little doubt that at this pace exhaustion would have been hastened, and the more so as chalking, marling and other permanent improvements were falling into disuse through the insecurity of the tenant's position.

A further improvement was soon to take place in soil fertility. A great advantage of the enclosure over the common fields was that crops could now be grown in autumn and winter; obviously this course was impossible when the village cattle strayed at will over the land from Lammastide to Candlemas. Consequently about the middle of the seventeenth century turnips, clovers, and cultivated grasses came in from Holland—the source of many of our great improvements—and slowly took their place in our agricultural system. Before these new crops could be cultivated a great improvement was needed in methods of tillage. The old implements were very crude: heavy wooden ploughs turned up the earth in great clods that could not be broken up by the inefficient harrows; so that after the seed had been broadcasted the clods had to be broken in pieces by large wooden hammers. "It is a greate labour and payne to the oxen, to goo to harowe; for they were better to goo to the plowe two dayes, thanne to harowe one daye. It is an old saying, 'The oxe is neuer wo, tyll he to the harowe goo.'...And if the barleye gounde wyll not break with Harrows, but be clotty, it wolde be braken with malles, and not streyght downe: for than they brake the corne in-to the earthe," wrote Fitzherbert in 1543. Two hundred years later Tull declaims against farmers who, "when they have thrown in their seed, go over it twenty times with the harrows, until the horses have trodden it almost as hard as a highway." The young plants thus had considerable difficulty in getting through, and later on in the season they were terribly hampered by the excessive growth of weeds which could never be got rid of by the old methods.

These defects were only slowly remedied, but the man who probably did more than anyone else in this direction was Jethro Tull. Travelling in the South of France and in Italy in the early years of the eighteenth century, he observed how carefully the vineyards were cultivated. On his return home he adopted similar methods on his farm at Shalbourne, on the borders of Berkshire and Wiltshire, adapting and inventing the necessary implements. Some farmers, indeed, had already begun to get a fine tilth : he tells us of "Great quantities of very light land (in Gloucestershire) which when kept in the *sat erit*[1] husbandry were let for half a crown an acre, but being now brought into the pulverising method, are let for ten shillings an acre. But there is a misfortune in many parishes, that the custom does not permit any one to pulverise his light lands by tillage, until an enclosure be made of them."

Tull insisted on three points: (1) that the soil must be thoroughly pulverised before the seed is sown, (2) cultivation must continue after the seed is

[1] Tull's name for the old style.

sown and as long as is practicable, (3) the seed must therefore be sown in straight lines and not scattered broadcast. The seed drill and horse hoe that he made to carry these principles into practice have been the forerunners of a long line of useful implements, and they alone made possible the cultivation of turnips, clovers and other of the new crops that began to come in.

These new crops completely changed the agriculture of the country. They fell best into the rotation worked out by Lord Townshend in the middle of the eighteenth century: clover in place of the old fallow, then wheat, then turnips, and lastly barley. Three great advantages followed. The clovers and other leguminous plants, sainfoin, lucerne, etc., led to a great increase in the stock of soil nitrogen. The substitution of a growing crop for the fallow considerably reduced the wastage of plant food through leaching. The clover hay and turnips provided food for the live stock throughout the winter, so that it was no longer necessary to slaughter them in late autumn and salt them down to be eaten[1]; the husbandman was able to keep them alive and in good condition all the year round. The animals consumed

[1]   "At Hallowtide, slaughter time entereth in,
       and then doth the husbandman's feasting begin."
                                        Tusser, 1573.

a great deal of food, but this did not necessitate a great net loss to the soil. For the carbon, hydrogen and oxygen (the elements most largely retained in the body tissues) came from the inexhaustible supplies in the atmosphere, while of the other constituents, retained to a less extent and largely passed off into the manure, the nitrogen was mainly drawn from the atmosphere through the agency of the clover, and only the phosphorus, potassium, calcium, etc., came exclusively from the soil. This manure when put on to the land actually enriched it in nitrogenous organic matter, and went far to replace the mineral substances withdrawn by the previous year's crops. So far as nitrogen was concerned, the system was permanent: crop production went on at a higher level than was ever before possible, and this new level was determined in principle by the amount of nitrogen fixed by the quadrennial clover crop, and in practice by the amount that was returned to the soil in the manure.

But there still remained the loss of phosphorus, which was intensified as the cities grew and imported more and more dairy produce, meat, bread, etc., from the country. At the end of the eighteenth century, in spite of all the improvements, an ordinary yield of wheat was probably only about 23 bushels, little more than good farmers had been accustomed to get for 250 years past. The improvements had been by no means universally adopted, and in many districts medieval agriculture was still the rule. It is impossible to determine how far the low yield was due to this circumstance. But in many cases the small supply of phosphates in the soil was now the limiting factor, preventing the crop from rising.

Then gradually bones came into use as manure and produced such remarkable results that from the early years of the nineteenth century considerable quantities were imported from Europe. It would, perhaps, be unkind to enquire too particularly where they came from: Liebig roundly declared that "England is robbing all other countries of their fertility. Already in her eagerness for bones she has turned up the battlefields of Leipsic, and Waterloo, and of the Crimea: already from the catacombs of Sicily she has carried away the skeletons of many successive generations....Like a vampire she hangs upon the neck of Europe, nay, of the whole world, and sucks the heart blood from nations without a thought of justice towards them, without a shadow of lasting advantage to herself."

But even finely ground bones sometimes acted only slowly and sometimes failed to act at all. This was the case at Rothamsted: during the years 1836—1838 Lawes had used bone dust on turnips without avail, although it was effective elsewhere. He therefore prepared the soluble calcium phosphate, then known as superphosphate, by treating the bone with sulphuric or other acids. This proved remarkably effective on turnips; and he took out a patent in 1842 and commenced the manufacture on a large scale. But as the source of the calcium phosphate was immaterial, he used mineral phosphates and guanos instead of bones.

Another source of phosphatic manures was opened up in 1879 when Thomas and Gilchrist introduced their process for removing the phosphorus from iron during its conversion into steel. At first the agricultural value of the basic slag thus produced was not recognised, but it was slowly revealed by the experiments of Wrightson and Munroe in 1885, and of other agriculturalists.

Phosphatic guanos, brought from the Pacific Islands, fish guano worked up from refuse fish, and meat guano from imported meat, contribute in a lesser degree to the farmers' supply.

As a result of having these large supplies of phosphates from various parts of the world, farmers now very generally add phosphates to their land, and thus remove the limiting factor which had in many cases kept down the crops. Concurrently with the increased use of phosphates there has been a marked increase in soil fertility, the yields of turnips in particular have gone up very much, and there has been very great improvements in the pastures. More cattle can therefore be kept and more manure can be made for the arable land.

It often happens, especially on the lighter soils, that the crop supplied with phosphates is now limited by the deficiency of potassium salts. This deficiency has long been partially met by dressings of wood ashes, salt, etc., but a better method was needed. Fortunately large supplies of potassium salts were discovered at Strassfurt in Germany and were put on the market in 1861. Since then they have been extensively used, although curiously enough no similar deposits have been found elsewhere.

The modern farmer is no longer dependent on leguminous plants for his supplies of nitrogen. Nitrate of soda is imported from Chili, sulphate of ammonia is manufactured from coal at home, and a large number of grains and seeds are imported from over the seas to feed the cattle and thus increase the supply of farmyard manure.

Modern agricultural systems are far too complex to reduce to any simple rigid order ; but their general bearing on the fertility of the soil may be briefly summed up:

1. The supply of plant nutrients is kept up by the addition of appropriate artificial manures. It is impossible to determine *à priori* either by chemical analysis or otherwise exactly what mixture of manures will be best; nothing but direct trials suffice. But a

number of trials are being made, and in some cases
on a definite systematic plan, to ascertain broadly the
needs of the commoner crops on important types of
soil. Chalk or lime is also applied, though, it must
be said, not always as often as necessary.

2. Every effort is made to keep up the supply of
nitrogenous organic matter in the soil. It is not yet
possible, and perhaps never will be, to maintain the
same level as in land permanently covered with grass
or other vegetation. But leguminous crops are grown;
the "seeds" (i.e. mixture of grass and clover) are left
for two or three years, in which time a dense root
mass forms; and in many instances a crop (such as
mustard, tares, etc.) is sown with the deliberate in-
tention of being ploughed into the ground.

3. Enormous quantities of cattle food are imported from newer countries and from less highly
developed regions. Only a small part—not more
than five or ten per cent.—of their nitrogen, phosphorus and potassium is retained by the animal: the
rest passes into the manure and goes to fertilise the
land. There are, however, very considerable losses
in making manure, and as much as one half of the
fertilising constituents may fail to reach the soil.
Greater economy is effected by feeding the animals
on the arable land, thus obviating the necessity for
making farmyard manure.

4. Crops are therefore grown suitable for animals

to eat in the field (sheep are the most convenient for the purpose). These crops include swedes, rape, kohl-rabi, thousand-headed kale, mustard and the aftermath of clovers and cultivated grasses[1]. The purchased food is supplied in troughs and the animals are confined by hurdles to a particular area till the crop and sufficient additional food are consumed. Then they are moved on, till finally they have covered the whole field.

5. Considerable tracts of land are laid down to permanent pasture. Judicious management of grazing, combined with dressings of basic slag, lime and potassium salts if necessary, lead to the formation of a dense turf of grass and clover. The land thus gains considerable supplies of nitrogenous organic matter and its fertility rises to the upper equilibrium level. These areas of well-managed grass land constitute, perhaps, the most fertile soils we have, and their fertility is more permanent, and maintained at lower cost, than that of any other soils.

Thus the modern English farmer keeps up the fertility of his soil by importing phosphates from the United States, Tunis, Algeria, Belgium and France; nitrates from Chili; potassium salts from Germany. He also imports grain—maize, wheat, barley, oil seeds, etc.—rich in valuable fertilising materials from the United States, Russia, Roumania, Argentina,

---

[1] Often called "rotation grasses."

British East Indies, Canada and other parts of the Empire. Thus a prodigious transfer of soil fertility is taking place from these countries to our own. The process at present is enormously wasteful. We have seen that terrible losses of fertility may arise under conditions of pioneer farming; even the remnant saved in the crop suffers further loss in many a badly arranged British farmyard and exposed manure heap.

Lastly, the crops raised on the British farm are largely sent off to the cities from whence only little manure ever returns, the great proportion of the fertilising constituents getting into the sewage and being destroyed at considerable expense.

It is obvious that such wasteful methods cannot go on indefinitely. Investigations into the losses and gains are now going on at Rothamsted and elsewhere. With fuller knowledge there is little doubt that some of the waste can be reduced, while the action of the recuperative agencies in the soil can be accelerated. It is impossible to overestimate the importance of evolving a permanent system of maintaining soil fertility, but such a system must rest on a solid foundation of scientific fact.

CHAPTER V

THE RAISING OF THE FERTILITY LIMIT

WE have seen in the previous chapters that the fertility of a given soil may lie anywhere between two limits: the higher limit being attained when the land is allowed to remain with an undisturbed vegetation of grass and clovers, and the lower when the land is perpetually under the plough, producing nothing but cereal crops and receiving no manure to counterbalance the various losses. We have further seen how, by a judicious system of husbandry, it is possible to maintain arable land somewhere near the higher fertility limit by arranging for recuperative periods in grass and clover, and systematically adding manurial substances to replace whatever may be lost.

The higher limit beyond which the fertility of the soil as it stands cannot be pushed, is set by the nature of the soil, its position in respect to water supply, climate, etc. But it is often possible to change these, and when this is done the fertility is no longer tied down to the old limit but rises to a new one set by the new conditions. This process is of course fundamentally different from the case we have just dealt with, and is in practice so much more costly that the

Fig. 3. Land left to run wild at Rothamsted. After 25 years it becomes a dense thicket

distinction is recognised both by custom and by law: the maintenance of fertility between the natural limits is regarded as the tenant's business, while the extension of the upper limit is considered to be the landlord's duty.

Over the greater part of England such an extension of the fertility limit has taken place. Often indeed the extension was necessary before the adoption of expensive modern methods of farming could be justified. The process has generally involved some conflict with natural conditions, and the new order of things stands only so long as constant care is exercised. In most cases the original condition is resumed if the intervention of man ceases for a time; a plot of land at Rothamsted that has been left to itself since 1882 is now a dense thicket and bids fair to become an impenetrable wood before long. But the expenditure necessary to maintain the new limit is much less than that required to reach it, so that in this sense the improvement is entitled to be called permanent.

Some of the land of England has always been open grass-covered Down land, and this was inhabited even in prehistoric times. Much of the land, however, was covered with forest which had to be cleared away before fields could be made but which would, as the Rothamsted experiment shows, soon spring up again if the suppressing hand of man were removed. It is difficult for us now to realise the magnitude of the task which medieval man set himself in clearing the forest with his imperfect tools and the enormous amount of labour that must have been required. Even with modern appliances—explosives and well-constructed jacks—the task is considerable, and the traveller round the shores of Lake Erie can still find many fields from which the timber has been removed, but the stumps still left, because the labour of removing them is so great that no adequate return could be obtained.

Clearance was still going on in England even as late as the middle of last century and an interesting account of one is preserved in the *Journal of the Royal Agricultural Society* for 1863. About ten miles west of Woodstock lies the forest of Wychwood, that formerly occupied considerably more land than it does now. In 1853 an Act was passed permitting disafforestation, and in October 1856 the work was begun. The account deals only with the portion allotted to the Crown, an area of nearly 3000 acres lying in the triangle between Fulbrook, Field Assarts, and Shorthampton. Of this nearly 2000 acres were "unreclaimed forest land, dense, dark, and gloomy: its silence seldom disturbed, except by the axe of the woodman, the gun of the gamekeeper, or the stealthy tread of the deer stealer."

Ten miles of road were first made, and these, with their boundary walls, cost £6985. Then it was necessary to get rid of the deer. "The Commissioners' order had gone forth against the deer 'let not one remain.' Some few were caught alive in nets, and taken away to stock distant parks, but by far the greater number had to be killed, and to effect this purpose the keepers were fully employed; to assist in the slaughter, guns and gunners came from the surrounding neighbourhood....As a complete clearance was to be made, bucks, does, and fawns, in season and out of season, shared the same fate, and the taste of venison was known in cottage as well as hall." Next the trees were cut down. "Hundreds and hundreds of men and boys were engaged, some cutting the light wood and laying it in drift, some tying the firewood into faggots, some preparing the larger pieces for posts and fencing and others busy felling the timber trees, or stripping off the bark." Some of the smaller trees were pulled down by a windlass worked by two horses. The total cost of this was £7742, but sales of timber, bark, etc. realised £21,823, and as £2450 worth was left untouched the gain on this part of the operation was £16,531. Next came the laborious process of digging out the roots known by the old Saxon name "grubbing"; this was accomplished by hand labour at a cost of £6233 for an area of 1903 acres. "Some of the roots were carried away to serve as fuel for the cottages near; but great quantities were burned on the land, rough firewood in the district having become so abundant, that it was not considered worth the expense of cartage." The ground being now clear, seven farms were measured out and whitethorn quicks planted along the boundaries ; houses and buildings were put up at a cost of £14,337. Allowing for all sales, the net outlay apart from the ten miles of road was £10,452 and the total farm land obtained was 2843 acres ; this was let at £5104 per annum, a gain of £3291 on the revenue derived from the forest. But the land was by no means ready for cultivation. The tenants, as they came into possession on 31 years leases, found "anything but a smooth, inviting appearance : wide ditches, and long, irregular high banks that had formed the boundaries of the different coppices ; deep pits and hollows, where stones had been dug for the use of bygone generations: small straggling briars that had escaped the notice of the woodgrubbers ; roots of trees and underwood, left a few inches below the surface by oversight or intentional neglect on the part of dishonest workmen; large patches of rough brown fern-stems, that had afforded covert to the fawns; all these and many other impediments stood in the way...it was with the greatest difficulty that four strong horses drawing a large iron plough could break up half an acre a day; and many and long were the blacksmith's bills for repairs. Some of the tenants tried digging, at a cost of £3 per

acre; some used stocking-hoes, and *grubbed* the
ground 5 inches deep, carefully picking out the large
stones that were beneath the surface; this plan cost
50/- per acre. On Potter's Hill farm, breast ploughing and burning was adopted; and this course
appeared to answer better than any of the others."
The banks were either thrown down by spade labour
to fill up the hollows, or gradually ploughed down.
Vast quantities of wood ashes were available for
manure and were spread as required on the fields.
Oats and turnips were grown during the first year,
much of the cultivation being done by hand; the
yield of the former was well up to the average of the
district and of the latter well above it (probably
because this crop had received superphosphate) and
for the five years over which the record extends the
tenant was satisfied with the returns. The landlord
(the Crown) was also well satisfied[1]. The recorder
further lays stress on the great moral gain to the
district—a point strongly emphasized by all advocates
of enclosure. "Formerly, when deer and game
abounded on the coverts, deer-stealers and poachers,
idlers, and thieves, were numerous around; conflicts
between them and the keepers were frequent; im-
prisonment and transportation caused many families
to lose their paternal head, and where matters did

[1] I understand that over the whole fifty years the returns have not
been so satisfactory.

not reach this point, perhaps the abiding influences were still worse, a stolen buck could readily be disposed of; the amount paid for such plunder frequently amounted to £2 or £3, but as ill-gotten booty is seldom well spent, the beer-shops too often absorbed the greater part of the proceeds. There was squandered in dissipation, what had been dishonestly obtained, a deserted home, a neglected wife, and children left to their own devices, fill up the background of this sad picture."

In this particular instance clearing only was carried out, but in many other cases further operations have been performed. Chief among these is drainage, which has been resorted to in all parts of England owing to the circumstance that the wetness of many soils more than anything else set the fertility limits and often in fact rendered them absolutely sterile. The old method in this country consisted in throwing the land into high ridges with deep furrows between, such as can still be traced in almost any clay district. Considerable waste of land was thus entailed: the furrows were often so wet that they lay bare of crop, whilst only the higher parts of the land were productive. No advance seems to have been possible in the common arable lands, as nothing could be done without the consent of all the owners; but on the enclosed fields better methods could be adopted.

A special drainage problem had, from time

immemorial, been solved successfully in various parts of England. Much of Romney Marsh in the South of Kent was reclaimed from the sea before or during Roman times, while the adjoining Walland and Denge Marshes were brought in and drained by successive Archbishops of Canterbury beginning about 774. The great monasteries in the Fens had also reclaimed parts of the surrounding land. In 1626 a great scheme was set into operation for draining the Fens and embanking its rivers, the work being executed by a celebrated Dutchman, Cornelius Vermuyden. As this is essentially an engineering problem we cannot go into the details of the methods adopted; nor does space allow any account of the romantic story of the project, its interruption by storm, by the exhaustion of the resources of the Adventurers, by the Civil War, and finally by the fenmen themselves, who had no taste for farming and no wish to see wheat and cattle take the place of fish and waterfowl.

"Behold the great design, which they do now determine,
  Will make our bodies pine, a prey to crows and vermin ;
  For they do mean all fens to drain and waters overmaster
  All will be dry and we must die, 'cause Essex calves want
      pasture"

went the old fenman's song.

The scheme as it works to-day consists of two great parts: (1) the water from the high lands is intercepted and discharged into the river so that it shall not reach the low-lying lands, (2) the water inside the low-lying area is drained into ditches and pumped into the river. To effect the first purpose a catch-water drain is cut at a point just above flood level and arranged to discharge by gravity into the nearest river. The second purpose is achieved by making open drains inside the catch-water and bringing them to the most convenient point for discharge. The water collected in these is below the level of the river and will not naturally flow in, but has to be lifted there. This is done by large scoop wheels, which are simply under-shot water-wheels driven the reverse way. In olden times power was furnished by windmills, and now-a-days by the less picturesque beam engine. Oil engines and centrifugal pumps are often used in new work.

The reclamation of Whittlesea Mere near Peterborough (Holme Station) affords an interesting instance of this type of problem. The Mere itself covered 1000 acres, and around it lay another 2000 acres of wet land or shoals. In 1844 an Act was obtained to improve the drainage of the Middle Level; a new cut 11 miles long was commenced to discharge the waters some six miles further down the Ouse than before, and so effect a lowering of the water table by six feet. Simultaneously connection was made with the Mere. In the summer of 1851 the new cut was sufficiently advanced to carry off the

Fig. 4. Reclaimed fenland, Lincolnshire

waters, the last bank was cut through and the Mere began to empty itself. There was a total fall of only two feet from the bottom of the lake and accordingly the stream was never rapid after the first twenty-four hours and was still flowing sluggishly even after three weeks. Fortunately a favourable wind prevailed and assisted materially the movement of the water. "Long before the last pools of water had disappeared from off the bed of the Mere," wrote Mr Wells, from whose description in the *Journal of the Royal Agricultural Society* for 1860 this account is taken, "large crowds of people from all the surrounding neighbourhood had assembled. Some from a desire to be present at the last moment of a venerable friend whose fortunes were now reduced to the lowest ebb: others perhaps with whom the love of stewed eels preponderated over sentiment, from the prospect of a ready and abundant gratification of their taste....Nine out of ten came provided with sacks and baskets to carry off their share of the vast number of fish, which, wherever the eye turned, were floundering in the ever-decreasing water. Some more ambitious speculators brought their carts, and gathering the fish by the ton weight, despatched them for sale to Birmingham and Manchester."

A pumping engine was now installed to carry the water table sufficiently below the surface for crop production. During the summer of 1852 the great expanse of mud was surveyed, farm boundaries marked out, and arrangements made for letting some of the prospective farms, when on November 12th the bank broke and the whole Mere was flooded again to a depth of 2½ feet. But the bank was mended and the engine set to work; in little more than three weeks the mud surface was once more exposed. Then a main dyke was cut through the area, and a number of smaller lateral dykes; this work was very arduous and the mud frequently fell in. But it was finished at last and the pump removed the water as fast as it collected. "The effect of this network of drains was quickly visible. The bed of the Mere was soon covered with innumerable cracks and fissures, deep and wide, so as to make it a matter of no small difficulty to walk along the surface, while in the surrounding bog the principal effect was the speedy consolidation of its crust....

"It was no easy matter to reduce the Mere-land into a state to receive such seed as should be first sown; the adhesive condition of the surface making it impossible to use horses even when shod with boards, if indeed the wide fissures did not render it dangerous to try the experiment. The whole area therefore had to be prepared by hand—over the largest part light harrows were first drawn by hand— the seed was then sown, and the harrows used a second and sometimes a third time, at a cost of about 5/- or

6/- per acre.   Other parts were dug or forked at an
average cost of from 25/- to 30/- per acre.   Of such a
depth were the cracks, that even this process with all
the subsequent operations attending the first crop,
by no means got rid of these obstinate scars, which
continued until the cultivation of three or four years
at length obliterated them."

Coleseed and Italian rye grass were the first crops
taken, and after that wheat and oats could be grown
owing to the richness of the soil and its large content
of calcium carbonate.   Excellent yields of these
cereals, of mangolds, potatoes and carrots were
raised.

There remained the more difficult business of
rendering fit for cultivation the tract of peat land
surrounding part of the old Mere.   This was done by
covering the peat to a depth of 4 to 6 inches with
the marl dug out of the dykes; the operation cost
£15 to £19 per acre but proved remunerative as the
land readily let at 30s. per acre.

Thus a vast unhealthy waste of marsh and mere
was transformed into healthy agricultural land and
made to produce food valued then at over £12,000
per annum.   To this day it remains a fertile tract.
One interesting change has, however, set in.   As the
water drained away so the soil shrunk, and it has
fallen in level to a remarkable extent.   Oak posts
driven into the underlying gault till their tops were

flush with the mud in 1851 are now more than ten feet above the surface.

Many of the magnificent alluvial meadows of the country have been made in the same way from rushy wastes. The well-known Brooks at Lewes give an example: originally only a bog of bullrushes, let for a trifling sum to chair bottom makers, they have for the past 80 years been fertile pastures carrying sheep and bullocks, yielding heavy crops of hay and contributing much to the wealth of the district.

These large schemes were early imitated by a few progressive agriculturists troubled with marshy or boggy fields. Walter Blith, a Yorkshire Puritan and "lover of Ingenuity," as he styled himself on the title-page of his *English Improver* (1649), had indeed already published methods that the farmer could adopt. He begins by pointing out that the causes of soil infertility "are usually two, 1 in Man himself, 2 in the Land itself. In Man himself it was occasionally, who by his sin procured a curse upon the Land, even Barrennesse." Of the defects in the land, one of the worst can be removed by "Drayning, or taking away Superfluous and Venomous Water, which lyeth in the Earth, and much occasioneth Bogginess, Miriness, Rushes, Flags, and other filth, and is indeed the chief cause of Barrenness in any land of this nature." He goes on to set out the essential condition that the drains must fall gradually but consistently from the

Fig. 5. Taking levels for draining in Puritan times
(From Walter Blith, *The English Improver Improved or the Survey
of Husbandry Surveyed*, 1652)

highest land to the lowest where the outfall must be, a condition which, even in the middle of the nineteenth century, was not always acted upon. "Be sure thy Drains be such, and so deep, as thou hast a descent in the end thereof to take away all thy water from thy Drayn to the very bottom, or else it is to no use at all, for suppose thou make thy Drain as high as an house, and canst not take thy water from it, thy work is lost; for look how low soever is thy lowest level in thy Drain, thou mayst drain thy water so low, and not one haire's breadth lower will it drain thy ground than thou hast a fall or descent to take it cleanly from thy Drain; therefore be especially carefull herein, and then if thou canst get a low descent from thence, carry thy Drain upon thy Levell untill thou art assuredly got under that moysture, miriness, or water, that either offends thy Bog, or covers thy Land; and goe one Spades graft deeper...to the bottom where the spewing spring lyeth thou must goe."

The drains must be laid out straight with as few "Angles, Crookes, and Turnings" as possible, and proper levels taken; the various tools and appliances needed are described and pictured in detail. Good green faggots of willow, alder, elm or thorn or else "great Pibble stones or Flint stones" are to be put into the trenches, on top of this some turf facing downwards, and then the whole filled in.

But it was nearly two hundred years before the plan was adopted on any wide scale by farmers or landowners. In 1823 James Smith, of Deanston, Perthshire, drained a marshy piece of ground in this manner and converted it into a garden. The interest of farmers in the experiment was aroused and maintained: in 1831 he set out the results of this and other trials in his *Remarks on Thorough drainage and Deep ploughing*. He recommended stone drains (like Blith's) 2 to $2\frac{1}{2}$ ft. deep to be made in the furrows, or, on flat land, 10 to 15 ft. apart in heavy soils but at wider intervals in lighter soils. Josiah Parkes, the drainer of Chat Moss, took a different view which he defended in his papers in the *Journal of the Royal Agricultural Society* (1846, etc.) and in his book *Philosophy and Art of Land Drainage* (1848). He maintains that drains should be deep—not less than 4 to 6 feet but they could be placed at wider intervals. The stones were soon displaced by John Reade's pipes of 1843, which in 1845 were turned out by the thousand in Thomas Scragg's machine. Throughout the 'forties and succeeding years drainage became a very popular improvement; public loans were raised, companies were started, and individuals expended their resources in developing great schemes. But the question of the depth of the drains was not settled; considerable controversy went on between the advocates of Smith's and of Parke's methods;

and it was not till thousands of acres had been wrongly drained and thousands of pounds wasted, that the germ of truth underlying both sides was discovered. For both sides were partially right; deep drains are needed to carry off subterranean water and shallow drains to remove surface water. Modern practice tends to revert to Smith's method; the drains are now commonly put $2\frac{1}{2}$ to $3\frac{1}{2}$ feet deep and 15 to 30 feet apart. In one other point a change has been made; the pipes are now often 3 inches in diameter instead of one or two inches as formerly.

Where the drainage was carried out effectively a most striking improvement resulted. The ground lost its wet sticky character. It could be ploughed earlier in the year, so that the seed could be sown early, and the crops safely left growing later. As the excess of water was removed air took its place; better root growth now became possible and considerable increases in crop were obtained. Great improvement also set in on the grass land; the reeds and rushes disappeared and the grasses and clovers flourished. But the change is not entirely permanent; the drains gradually become blocked up with silt, with a deposit formed of oxide of iron together with organic matter, and with roots of trees or plants; and considerable areas of land in the country now require redraining.

We have seen that the most generally fertile

soils are the loams, which consist of sand, silt and clay together with calcium carbonate. Soils lacking any of these constituents are usually less productive, but the fertility limits are raised directly the lacking constituent is supplied.

Thus a sandy waste may be made productive after addition of clay: and a dense clay soil may be ameliorated by adding calcium carbonate, and to a less extent by adding sand. These processes are simple enough in principle but require considerable human labour in practice, so that now-a-days they are relatively costly. They are among the earliest improvements in our agriculture, being known and practised by the Britons according to Pliny[1]. His account of the process affords an interesting glimpse of the agriculture of those far-off days. "The peoples of Britain and Gaul have discovered another method for nourishing the land. There is something they call marl (*marga*). It contains a more condensed richness, a sort of fatness of the land....There were formerly two kinds only, but several have lately been put to use by clever men: the white, the red, the columbina, argillaceous, tufa-like, and sandy are all used now. Marl is two-fold in nature: hard or fatty; these can be distinguished by the touch. In like manner it has a two-fold use; some kinds are used for crops (*fruges*) only, others for herbage (*pabulum*)

[1] Book 17, 6.

only. The tufa-like variety is good for crops. The white kind found in streams is extremely rich; it is hard to the touch; if too much is put on it burns the soil. The red kind is called *acaunumarga*, it contains hard lumps of petrified sandy fragments. This is broken on the field itself and in the first years the stubble is only cut with difficulty on account of the stones. It is put on very sparsely, only half as much being used as of the other kinds. They think it is mixed with salt. Either of these kinds put on the land will last for 50 years.

"Of the fatty marls the chief are the white varieties. There are several of these: the sharpest is the one above mentioned. Another is the silvery chalk. It is sought for deep in the ground, wells being frequently sunk 100 ft. deep with the mouth narrow and the shaft widening out as in mines. This is the kind most used in Britain. It lasts for 80 years and there is no instance of anyone who has put it on twice in his life time. A third white marl is called *glisomarga*, it is a fullers chalk mixed with rich soil more productive for herbage than for crops, so that after the harvest and before the next sowing there springs up a rank growth to be cut. When it is applied to crops it produces no other vegetation. It lasts for 30 years; if put on too thickly it strangles the soil. The columbina marl is called *eglecopala* by the Gauls, it is equally fertile. It is turned up in

lumps like stones and shatters into little fragments
under sun and frost. They use the sandy kind if no
other is available, and in any case on marshy soil.
The Ubii are the only people I know of who make
land fertile by digging up the soil to a depth of about
3 feet and then throwing on top a foot's thickness.
That kind does not last more than 10 years."

To this day the "silvery chalk" is dug out in
Hertfordshire just as Pliny describes: a well is sunk
and widened out to a chamber as the chalk is reached:
the chalk is hauled up and spread on the ground.
The well is partly filled in and leaves one of those
dells so characteristic of the fields of the county.

The method that Pliny ascribes to the Ubii has
been very much used. Light barren sands are often
underlain by heavier loam or clay which when brought
to the surface give rise to a fertile soil. Instances
will be given in Chap. VII.

It is not always necessary that the material should
be carried by human labour; the forces of Nature
can sometimes be utilised. Perhaps the best illustra-
tion is furnished by warping, a process introduced
during the early eighteenth century into North Lincolnshire and South East Yorkshire and practised
under the name of Colmatage in Tuscany, Romagna,
and the neighbourhood of Naples.

The Ouse, the Trent and other rivers connected
with the Humber are tidal, and as the flood travels

Fig. 6. Chalking in Hertfordshire. A well is sunk to reach the chalk, which is then hauled out

up stream it is seen to be loaded with mud scoured off from the banks and shores lower down. Much of the land lies below high-water level and consists of barren sand or peat. It is therefore divided up into areas of suitable size (200 acres or more is not uncommon) which are surrounded by banks and then connected with the river by means of wide channels fitted with sluice gates. When the flood is high the sluice gates are opened and the water runs over the area and is left to stand for 3 or 4 hours. There it deposits its mud: it is then allowed to run off and the mud is left to dry as much as possible before the next tide is due. The process is repeated daily at both tides so long as the tides are high enough; the inlets are periodically shifted and the incoming flood is skilfully managed to ensure that the deposit is spread fairly uniformly. In course of three years the deposit is some 2 or 3 feet in depth. The new land is now left undisturbed to dry for a time and to get some of its salt washed out; it is sown with white clover and left for some time to consolidate; then it can be drained and levelled and used for ordinary agricultural crops. Potatoes, wheat, roots, clover and rye-grass are commonly grown; they are often arranged in a three-year rotation, first potatoes (well manured with superphosphate and nitrate of soda, but no potash, as this is unnecessary), then wheat and finally roots, the clover mixture or oats. The land is very fertile, yielding 10 to 14 tons of potatoes, 7 to 9 quarters of wheat and still larger crops of oats; it lets readily at £2 per acre per annum and is considered to be worth £40 to £50 per acre, whilst the average cost of warping is only some £20. The improvement is permanent, although sometimes the shrinkage of the land becomes so great after a few years that re-warping is desirable to bring it up to its old level.

The process is obviously only possible where the land lies below the level of ordinary high tides. One large district, Thorne Moors, is in the main rather too high and an interesting modification is here adopted. The Moors are mainly peat, and peat is an article of distinct commercial value; it is therefore dug out, dried, and broken in a disintegrator; the coarser part is made up into bales and sold in the cities as peat moss litter, while the finer material is sent abroad to be soaked with molasses and then used as cattle food. Some also is distilled for the sake of its products. An area of about 200 acres is thus excavated to a sufficient depth—some 5 ft. or more—and the warping is then begun.

The defects of a clay soil cannot so easily be remedied as they arise from an excess of clay and fine silt rather than a deficiency of anything. Under special conditions reclamation has been effected by digging in great quantities of coarser material: a considerable tract of land is being treated with city refuse at Murieston, Midcalder, by the Edinburgh Distress Committee and a marked improvement in productiveness has resulted.

Thus the land that we cultivate to day is far removed from virgin land; it has been cleared, enclosed, levelled, often embanked, drained, chalked and marled by successive generations of cultivators. No small part of the difficulty of dealing with economic land problems arises from the great amount of capital that has been expended in the past in effecting the necessary improvements. In many cases the rent now received for agricultural land affords no adequate return for the outlay incurred even during the past sixty years. On the other hand it is arguable that improvements in land are a condition of national existence and therefore lie outside the scope of investments made for profit. We cannot now go into a discussion of these social and economic problems. The important conclusion is that our land owes much of its fertility to the labours of those who have gone before us. The improvements they effected are not wholly permanent but have to be maintained and renewed by each generation; any neglect of this duty may result in marked deterioration of the land and may necessitate considerable expenditure of time and money to bring back the fertility to the level at which it had formerly stood.

CHAPTER VI

THE CHEQUERED CAREER OF THE CLAYS

A CLAY soil needs no description. Everyone is familiar with the grey, green or dirty red coloured soil, sticky and slippery after rain, on which in winter time pools of water lie for days or weeks together. In the summer it dries to hard intractable clods, shrinking so much during the process that great gaping cracks appear, making the fields unsightly and in extreme cases even somewhat dangerous.

But although its general properties are very characteristic and easily recognisable no one has succeeded in drawing up any rigid definition of what is and what is not a clay soil. No sharp line of demarcation exists in Nature, and the clays shade off by imperceptible gradations into the wholly different class of soils known as the loams.

Agriculturally the clays are difficult to plough because of their stickiness, and for the same reason they make rather dangerous habitats for seedlings. It is no uncommon experience to have to sow a second time because the first lot of seeds have become asphyxiated. Even when the young plants have struggled through, their troubles are not at an end, for with the first spell of hot dry weather the soil dries to a solid crust that is little, if any, better for them than the wet sticky mass produced by heavy rain.

The great difference in agricultural value between clays and loams is sharply revealed by a study of the face of the country.

Loams, as we have already seen, are very generally fertile and are practically all under cultivation. In a loamy district almost every available piece of land has at one time or another been taken up, and little, if any, waste land is left. Space is economised as much as possible; there are few, if any, village greens or commons, and even the very lanes and roads are narrow and often worn deep by the heavy traffic of bygone days when road-making was still a lost art.

The hedges are straightened out and well kept, ditches are filled in unless actually wanted, and the whole country has a well-cared-for appearance. But in a heavy clay district there was less temptation to take in the land so completely. Indeed, some of the worst of the land probably never was taken in at all, but remains covered with forest apparently pretty much in its primeval state. Blean Forest near Canterbury, King's Wood running along the North Downs, many acres of wood in the Weald of Kent, all occupy land that has never been coveted by agriculturists, and so has always remained untouched. Even where the land has been taken up the process was very incomplete: village greens and commons have been left, the roads are wide, much wider, in fact, than need be now-a-days, so that only a part is made up and the rest is left as untidy picturesque wastes of bramble and briar, inhabited occasionally by a few roving gipsies or tramps, but of no practical value to anyone else. The fields are often small and the straggling hedges and ditches occupy a disproportionally large area of the land. The hedges are badly kept and the bushes have been allowed to develope into trees, so that looking over a clay region such as the Weald of Kent one gets the impression of a heavily wooded country. The farming is reduced to its simplest, grass only is grown because that involves least trouble and expense, and the land is worked with the lowest possible expenditure of money and labour.

Such heavy unremunerative soils can be found in places on the clays of the Coal Measures, the Oxford Clay, the Weald and elsewhere. They merge insensibly into lighter and more tractable soils, which in turn shade off into the fertile loams. But no limits can be set anywhere. At one end of the series we have valuable fertile loams, at the other end clay wastes; and somewhere in between come a number of soils which pay to cultivate when prices are high, but are unprofitable when prices fall; they then bring disaster on the holders and soon go out of cultivation. These soils occupy in the aggregate a pretty considerable area of the country, and their history is extraordinarily interesting, because it so accurately reflects the chequered life of the agricultural community.

Many of these borderland soils first came into cultivation during the Napoleonic wars when prices of wheat rose to the highest level ever reached. In the years of depression between 1813 and 1836 they went out of cultivation and were commonly allowed to cover themselves with weeds and grasses, and afford miserable grazing for unfortunate live stock. After they were drained in the early 'forties they became once more productive, and during the prosperous years of the 'fifties and 'sixties they were in great demand. Then when the bad times came, culminating in the disastrous season of 1879, the clay lands again fell out of cultivation. With a return of prosperity they were once again cultivated, and now-a-days we find them converted into good pasture for dairy cattle or fatting stock and into arable land growing wheat, mangolds, cabbage and, where possible, potatoes. It needs but little foresight to see that in the next wave of depression some of them may again go out of cultivation.

All this time, however, the loams have remained in cultivation in spite of all vicissitudes of prices and of seasons.

The story that we have briefly sketched out must now be studied with a little more detail.

The ancient method of dealing with clays was to lay them up in high backed ridges so that the rain could run off into the furrows. There it often lay for long periods. On these high backed lands cross-ploughing was impossible; cultivation was not deep; the surface being worked only to the depth of two or three inches and the subsoil was never touched. Only the ridge carried a crop of any size: the furrows were too wet in winter and too hard in summer to allow of plant growth.

Chalking and marling were commonly adopted in good times or whenever circumstances were propitious to permanent improvements, but they were neglected in bad times in spite of the advice of all agricultural writers. "Howsoeuer this Weald," writes Gervase Markham in 1625[1] of the clay plain forming the Weald of Kent, "be of itselfe vnfruitfull and of a barren nature, yet so it hath pleased the prouidence of the Almighty to temper the same, that by the benefit of Margle or Marle (as it is commonly called) it may be made not onely equall in fertility with the other grounds of the Shire, as well for Corne as

[1] *The Inrichment of the Weald of Kent*, 1625.

Fig. 7. Ridged land on the clay

Grasse, but also superiour to the more and greater part of the same." The antiquity of this process was demonstrated " by the innumerable Marle pits digged and spent so many yeeres past, that trees of 200 or 300 years old doe now grow upon them."

Paring and burning was also a common method of amelioration. The top two or three inches of soil were pared off, collected into heaps and burned. (The agricultural labourer has a wonderful facility for setting fire to the most unpromising material.) The process was so managed that the clay was not baked into brick, but sufficiently heated to disintegrate it and partially decompose the organic matter. Then the heated material was spread over the land and was found to be very productive. In some of the eastern counties Blith's draining method was early applied; thus Vancover in his *Report on the Agriculture of Essex* in 1795 refers to much of the stiff clay as being "hollow drained" and dressed with chalk, after which it continues to give good crops for 20 years or more. Wheat and beans have always been the most suitable crops for clay farming, and a common Essex rotation was: fallow, wheat, beans, wheat, or, where greater diversity was required, fallow, oats or barley, clover and rye-grass, wheat, beans. But the implements were cumbrous; in bad weather it might prove impossible to get the crops in, so that a season would sometimes be missed. So long, however, as wheat remained high in price occasional losses of this kind were not serious to the farmer. But the labourer suffered, for he found himself practically out of work all the winter, excepting when the ground happened to be frozen sufficiently hard to enable the dung carts to travel, or when hedging and ditching had to be done.

Next came a period of depression from 1813 to 1836 when much of the clay land became derelict.

A new era of prosperity opened with the reign of Queen Victoria and gradually the land was taken into cultivation. A delightful account has been preserved[1] of the reclamation during this period of a cold, wet, clay farm. "Every incoming tenant took it at about half the previous rent ; dabbled about for a year or two like a duck, and retired—'lame.' It was but a simple equation—a very simple one—to say when the rent would come to zero." The water did not drain away, "it would stand, day after day, and week after week, and month after month, shining along the serpentine furrows, as if it never, never, *never* would go again. And the only wonder was when or how, or by what bold amphibious being the ridges had ever been raised, which it intersected, like a sample series of Dutch canals and embankments."

[1] *Talpa, or The Chronicles of a Clay Farm*, by C. Wren Hoskyns, 1852.

A careful survey by the owner showed the apparently level area really had a fall of nine feet, so that systematic drainage was quite possible. This was clearly the first step to be taken. Calculations were made to show the amount of fall that must be obtained in each field, and the men were set to work to open up the trenches, so that the levels might be taken previous to laying in the pipes. With so small a fall it was necessary that the work should be accurately done.

But the whole idea was new to the labourers. They had never seen telescopes and levels and they were convinced that the farm was level and undrainable. "The morning after my head-drainer had commenced operations I found him hard at work cutting a drain about eighteen inches deep, *laying in the tiles one by one, and filling the earth in over them as he went!*...I began something in this way—'Why, my good friend, what on earth are you about? Didn't I tell you to lay the drain open from bottom to top, and that not a tile was to be put in till I had seen it and tried the levels?'...Every inch of depth was of value at the mouth of so long a drain. 'Three feet deep at the outlet' was the modest extent of my demand; and there I stood watching the tiles thrown in pêle-mêle to a depth of eighteen inches, which I was given to understand was 'about two feet' with as cool an indifference to the *other* foot, as if Two and Three had been recently determined by the common assent of mankind to mean the same thing.

"'But I *must* have it three feet deep.'

"'Oh, it's no use: it'll never drain so dip as that through this here clay.'

"'But I tell you it *must* be. There can be no fall without it.'

"'Well, I've been a-draining this forty year and I ought to know summut about it.'

"At that instant my eyes began to open to the true meaning of those 'practical difficulties' which the uninitiated laugh at, because they have never encountered them; and the man of science despises, who has said to steam, water and machinery 'do this,' and they do it, but has never known what it is to try and guide out of the old track, a *mind* that has run in the same rut 'this forty year and more'."

By a skilful appeal to the old man's vanity the matter was rectified and the drains properly laid. Then came the next improvement—throwing down the high ridges in which the land had formerly been laid, and flattening out the field, so that the implements could work more easily. This met with even more serious opposition, and a long struggle ensued with the collective experience of the district. "My own working bailiff headed the attack within the camp; while a neighbouring clergyman led on the foe from without, evidently viewing the heresy in a serious light, and myself as a fit subject for an *auto da fé*. The conclusion of our last skirmish was too good to be lost to posterity. I entered it *verbatim* in my farm memoranda.

" 'But tell me in earnest. Don't you mean to ridge up that field again ? '

" 'No.'

" 'What, you mean to lay it *Flat* ? '

" 'Yes.'

" 'In the name of Goodness, Why ? '

" 'Because *the name of Goodness*—made it so.'

" If I had suddenly assumed some demoniacal form, and then, leaving a train of smoke and brimstone, vanished, with a clap of thunder, from before the eyes of my catechist, I do not think his face would have assumed a greater expression of resourceless and complete astonishment[1]."

Next, the material thrown out from the drains was put on the surface of the land—an operation that was regarded as the crowning act of folly and brought up the wise men from far and near to look and scoff.

Lime was now put on.

[1] The worthy clergyman's astonishment was not wholly unreasonable because in levelling the ridges a considerable amount of very unkindly subsoil must have been exposed, which would only slowly weather down into a decent soil.

Then the small fields had to be made into bigger ones : hedges were grubbed up and the banks were thrown down, much to the disgust of the local fox hunters and rabbit shooters.

Turnips were then sown—they had never previously been grown on the farm—and they were fertilised with guano which was then just coming into the country. This evoked much comment from the local wits, but the crop was magnificent, being far the best in the countryside. "It was stared at and stared at again, as a sort of conjuror's trick which '*You couldn't do again*.' 'Wise men shook their heads and held their tongues at it. Nobody would have been at all surprised if, on going to the field some fine morning, he had found it altogether vanished, like faëry money, as quicklv as it came : and as the roots swelled and swelled into confirmed substance and reality through September and October, the silence about it became perfectly portentous.... Where did the crop come from ? How did it grow ?... Surely it must at any rate be but a fraud upon the land after all ; and the *next crop* would show the different results of *real manure* and a *mere stimulant*. This was the point to which *opinion* at last settled down. 'We'll wait and see' was the final determination expressed."

The big crop of turnips enabled sheep and cattle to be kept, and their manure helped to enrich the land and to keep the fertility up to the new level to which the drainage and liming had brought it. But the introduction of live stock had far-reaching economic effects also : it afforded employment for the labourers during the winter, and it had a steadying influence on the farm receipts. For when the price of wheat was low that of meat was high, and *vice versa*, a relationship that crystallised into the saying, "Up corn, down horn."

In this way many clay farms were made fruitful ; lime and chalk once more came into use, and the introduction of artificial manures and of concentrated feeding stuffs for the animals contributed largely to the increase in crop production that was taking place.

But the tide of prosperity began to turn, and in the late 'seventies a run of bad times set in, ruining many farmers and throwing out of cultivation much of the land that had been reclaimed. It lay for years neglected and covered with grass and weeds ; its only use was to afford a little poor grazing for live stock. It was certainly gaining fertility and increasing its stores of nitrogenous organic matter, but it afforded little sustenance to the farmer. Essex, which had in the 'sixties been extremely prosperous, looked like becoming derelict : other clay counties fared no better. Many of the farmers who survived met the crisis by laying down their land to grass, dismissing their labourers and reducing their working expenses to a minimum. Great was the distress all round. Tales of those days are still told in the villages, and are indeed often the only information possessed by the well-meaning agricultural reformers who dwell in the cities.

Part of the agricultural depression was due to the opening up of the western states of Canada, and part of the recovery was, in return, effected by the labour-saving machinery invented and made out there. Under the older system in vogue in 1890 the cost of harvesting wheat came to 27s. to 30s. per acre on a certain large corn farm : with the new binders the cost was reduced from 1897 onwards to 16s. to 18s. per acre on the same farm. Further, new methods began to come in. Essex was invaded by good Scotch farmers who were untrammelled with any views as to the necessity for growing wheat and beans, and turned instead to dairy produce and potatoes. For miles round London and other large cities dairying has saved the situation and once more brought into use land that had gone derelict. Elsewhere (e.g. in parts of Leicestershire) cattle are bought and fattened, while costs of production are cut down by the introduction of labour-saving devices and by skilful management. The introduction of the mangold into British agriculture has been a great boon to the clay farmer ; this crop is much more reliable than swedes on clay land and, together with cabbage, which also does well, affords valuable succulent food to the animals, while on the lighter fields of the farm swedes can also be grown. Much food has to be purchased—brewers' grains and cotton cake being special favourites—and this contributes to the fertility of the land. Large quantities of manure are also imported, for mangolds respond perhaps more than any other crop to liberal treatment, and are found to yield most profit when well manured. Thus the fertility of the arable land is being pushed well up. But the mainstay of clay farming is the grass land. Grass is the cheapest and easiest crop to raise and is steadily gaining ground at the expense of the arable crops. Temporary pastures figure very prominently, particularly in northern systems of agriculture. Magnificent permanent pastures are found on some of the better clays of Leicestershire, and on the low-lying alluvial flats round the estuaries of some of the rivers and elsewhere; some of these with very little trouble will carry and fatten live stock. Over large areas, however, the grassland is poor, but it is now receiving considerable attention. Although for years it often carried nothing more than a poor, thin growth of weeds and grass it really did not need any very great outlay to be considerably improved. A dressing of 10 cwt. of basic slag per acre has often a wonderful effect in increasing the growth of clover and producing a more nutritious herbage.

Trouble still arises from the presence of epizootic diseases in animals grazing on clay land, but this will no doubt be overcome as fuller knowledge is gained of the fauna of the soil. Often, however, the drainage is faulty. Over much of the Midland clay area the drains were laid in the middle of the last century at a depth of 4 ft.; this is now known to have been too far down. Here the trouble arises from the slowness with which the rain water gets away. The drains should, therefore, only be placed about $2\frac{1}{2}$ feet deep. In many cases also the grassland needs ploughing up and resowing with a suitable mixture. Finally, many clay farms need a good dressing of lime or chalk over the whole land, arable and grass alike.

Once the grassland is improved it commonly gets well treated; manure is put on if it is cut for hay, and concentrated food is supplied to the animals put out to graze on it. The grassland is consequently maintaining or increasing its fertility. At the present time, therefore, clays within reach of cities have distinct possibilities. Those further off, however, are frequently in a poor state and are more famous for the fox-hunting they afford than for their agriculture. Sometimes summer milk is produced to be sold to the cheese makers, but this trade is at a standstill in winter; sometimes also young store stock are raised to be sold off to the better farms. For it is a general rule that the raising of young store stock is most suitable to the man who farms poor unimproved land without much capital, while dairying and fattening are most suitable to the man who is going in for high farming.

The clays have probably never been managed on sounder lines than they are at present, but the lesson of history is absolutely clear; these soils are very apt to suffer in bad seasons and to ruin their occupiers in times of depression. A good margin must therefore be allowed for contingencies. Especially ought small holders and beginners to remember that the profit these soils can be shown to yield over a run of good seasons changes with disastrous suddenness to serious loss as soon as bad times come.

CHAPTER VII

THE RISE OF THE SANDS

STARTING once more from the fertile loams, a succession of soils can be traced, getting lighter and lighter and finally ending in the coarse material of heaths and sand-dunes. Thus we can begin with clay wastes, work through the fertile loams, pass on to cultivated sands and finally end in sand wastes, and find all the way gradual transitions with never a break to mark off the different classes of soils. A typical sandy soil is just as characteristic as a typical clay, but it equally defies rigid definition.

In many respects sand is the opposite of clay in general properties. Sandy soils have little power of holding water and therefore dry very readily; they do not long remain wet even after heavy rainfall. They are not sticky. The rock from which they are formed is generally somewhat hard, and so it has often happened that they have suffered less erosion than the clays, and have not, like the clays, been hollowed out into broad valleys. The ease with which water percolates through sand has led to the washing out of much of the soluble material, so that little is left except hard insoluble mineral grains which furnish but scanty food for plants.

Agriculturally the sands are a very mixed group. Their small power of retaining water is a serious disadvantage, partly because they become liable to drought, and partly also because of the ease with which manurial substances are washed away and lost. A good many of these soils happen to lie in relatively low situations and to receive underground water from the land above; these are often sufficiently supplied with water for all crop purposes. Others lie rather too high to enable the underground

water to be utilised. Thus the value of a sand
depends very much on its situation. A soil that
is fairly uniform may be fertile in one place where
water is available, but infertile in another not far
off, where the water is out of reach. Where cultivation is possible it is very easy : the land can be
worked almost directly after rain, seeds can (and in
fact must) be sown early in the year, and crops ripen
quickly and easily.

The sands that most resemble the loams—the so-
called sandy loams—have usually been in cultivation
as far back as any record goes. The lighter sands
have only slowly been taken up, and the process is
not yet complete, considerable areas being still left
as waste.

The stretch of country surrounding Fakenham and
Wells in Norfolk is classical ground for the student
of agricultural history. It was to Raynham, near
Fakenham, that Charles, second Viscount Townshend,
retired in 1730, after his political life was over, and
began those farming experiments that were destined
profoundly to influence our methods of husbandry.
At the outset the land was a barren sandy waste.
His first step was to apply marl which considerably
increased its productiveness. This, however, was no
new discovery ; marl was well known in Norfolk,
although it had long fallen into disuse. Lord
Townshend's great advance was the clearness with

which he recognised the conditions that make for fertility in sandy soils. Owing to their small retentive power they have to receive frequent dressings of manure, and this course is only possible where a considerable number of animals are kept. Means therefore had to be designed for combining animal husbandry with crop growing—two branches of farming which in the past had often been found mutually antagonistic. Lord Townshend's method was to grow turnips on the large scale, and then allow the animals to eat the crop *in situ*, so that their manure might fertilise the land for the next crop and their treading might consolidate it and so improve it as a seed bed. After turnips a crop of barley was taken and after this a crop of grass and clover, part of which could be cut as hay to supply food for the animals during winter, and the remainder eaten in the field by the animals in order to fertilise the ground for the wheat crop. Then turnips were taken again. The clover increased the stock of soil nitrogen and insured the permanency of the system so far as nitrogen is concerned. The plan was thoroughly sound and entirely successful; a manuring crop was taken, and then a cereal crop, then a second manuring crop and then another cereal crop. Both animals and crops flourished. So good is the plan that it survives to this day under the name of the Norfolk rotation, and many progressive farmers still use it with but the small modification that they often grow two corn crops in succession after the turnips.

But it commonly happens in the history of agriculture that improvements are adopted only very slowly, and Townshend's improvements were no exception to the rule. Certain difficulties also arose which Townshend did not overcome. Turnips were found to be liable to attacks of a minute beetle, *Phyllotreta nemorum*, commonly known as the Fly, which in dry weather sometimes almost destroyed the crop and left the animals without any food for the winter. Red clover (the ordinary variety grown) will not always grow every fourth year, but sometimes fails after the second or third course for some reason which is still obscure. Thus under the combined attacks of Turnip Fly and of Clover Sickness the farmer might find himself with a number of animals on his hands and no food for them, an awkward predicament from which he rarely extricated himself without considerable financial loss.

Fortunately another public-spirited landowner in the same district came forward and continued the experiments. Thomas William Coke, afterwards Earl of Leicester, inherited in 1776 his uncle's estate at Holkham, about twelve miles north of the scene of Lord Townshend's labours. The country was poor in the extreme. "All you will see," said old Lady Townshend to young Mrs Coke as she was going for

the first time to her new home, "will be one blade of
grass and two rabbits fighting for that." Coke's bio-
grapher and great-granddaughter, Mrs Sterling, thus
describes it[1]. "When Coke came into the property
the whole district round Holkham was little better
than a rabbit warren, varied by long tracts of shingle
and drifting sand, on which vegetation, other than
weeds, was impossible....Indeed throughout the county
of Norfolk the agriculture was of the poorest description. Between Holkham and Lynn not a single ear
of wheat was to be seen, and it was believed that not
one would grow. All the wheat consumed in the
county was imported from abroad. And, meanwhile,
everything that ignorance could do was done to
impoverish further an already miserable soil. The
course of cropping where the land would produce
anything was three white crops in succession, and
then broadcast turnips. No manure was ever purchased. The sheep were a wretched breed, and,
owing to the absence of fodder, no milch cows were
kept on any of the farms." Coke does not seem to
have begun experimental farming out of any abstract
desire for knowledge; he was led to it by the obstinacy
of an old-time farmer named Brett. The lease under
which this man held his farm had fallen in and was
under discussion for renewal; the original rent had
been eighteen pence per acre; this was subsequently

[1] *Coke of Norfolk and his friends.*   London, 1908.

raised to three and sixpence, and Coke now wanted five shillings. "Mr Brett jeered at the suggestion," continues Mrs Sterling, "and pointed out that the land was not worth the eighteen pence an acre originally paid for it. This was sufficient for a man of Coke's temperament, he immediately decided to farm the land himself."

No adequate history of Coke's agricultural work has been written[1], but from 1778, when the little incident just mentioned took place, down almost to the time of his death in 1842, he continued to make advances in the management of sandy land and disseminated his results at the great annual gatherings, the "sheep shearings," which for 43 years he held at Holkham. Realising the beneficial effects of grass and clover on the land, he left these crops growing for two, three, or even four years, thus adding to the nitrogenous organic matter of the soil, besides getting supplies of hay for the animals. Marl was applied in the first year at the rate of 80 to 100 loads per acre and left to wash into the land as long as the grasses stood there. When the land was ploughed up wheat was sown. Usually the amount of farmyard manure was insufficient for this crop, at any rate in the early days of the improvements, and manure had to be purchased. Rape cake (an old fertiliser in

[1] Some account is given in Dr Rigby's *Holkham, its Agriculture, etc.*, 1816. 3rd edition, enlarged, 1819.

Norfolk, brought in from Ireland and from the English mills) was therefore applied with excellent results: it cost £5 per ton and was used at the rate of 5 or 6 cwts. per acre: later on, however, the price rose considerably. After wheat, turnips were grown, and then barley followed by grasses as before. Thus the rotation was that of Lord Townshend except that the grasses were left growing for several years, and peas appear sometimes to have been sown after the grass was ploughed up and before the wheat crop was taken. Alongside of this improvement in cultivation he effected great improvements in the live stock of the district. He compelled his tenants to adopt a proper rotation and induced them to purchase good animals. So successful were his efforts that as early as 1784 Young states that "Mr Coke resides in the midst of the best husbandry in Norfolk, where the fields of every tenant are cultivated like gardens." There was a great surplus of produce, wheat and live stock were sold and the whole district became very prosperous.

The difficulties inherent in the Norfolk rotation—turnip fly and clover sickness—now engaged his attention. Although he did not surmount either difficulty (no one has done so even yet) he got over them by increasing his range of crops so that he should not be wholly dependent on turnips and clover. Instead of having the whole of the land in four crops he devoted some of it to other fodder plants. Sainfoin in particular proved valuable; it yielded considerable quantities of nutritious hay for winter, and, being a leguminous plant, it greatly enriched the soil in nitrogenous organic matter. Tares also were sown; some in October or November, some in April and May, to afford more green food to the animals.

Later on he grew mangolds, cocksfoot, potatoes, and he made experiments with other fodder crops. He purchased oil cakes for his animals, and thus not only fattened them more rapidly but also increased the amount of fertilising material in the manure. In this way he imported fertility from other districts to his own, a process which has now become a regular part of British husbandry. Thus sheep and cattle were the central feature of the farm as in Lord Townshend's system, but Coke increased the margin of safety by having certain areas of other fodder crops not liable to the same ills as clover and turnips, so that if one set of troubles intervened he would still have a reserve of food for his animals.

Little has been added to our knowledge of the best methods of farming sandy soils, and in all essentials our best present day methods are practically the same as these. The reclamation of the sands was now within the power of any landowner and was soon taken in hand in many districts. The Duke of

Bedford reclaimed much of the sand at Woburn, and before long the old parish turbary was waving with corn. An example of a later reclamation is afforded by Delamere Forest, Cheshire. The marl pits having been formed and opened, a tramway was laid from the pits to the land. Dressings of marl were given varying from 100 to 180 cubic yards per acre at a cost of £7 to £10; in consequence the land which before marling was not worth 5s. per acre afterwards let at £1. 10s. per acre[1]. The light sand of the Pays de Waes, lying between Antwerp and Ostend and traversed by the Waesland railway, has also been reclaimed by the application of clay or marl.

Sometimes, however, the barrenness of a sandy soil is due to a layer of rock or a "pan" lying near the surface and interfering so seriously with the movements of the soil water that proper plant growth cannot take place. In such cases the only possibility is to break up the rock and pick it out, a laborious enough process even now when steam implements are available, and still more so in the early days. An example is furnished by Coxheath, an area of some 900 acres near Maidstone. This used to be waste land, but in 1814 an Enclosure Act was obtained. The ground was then trenched and the layer of rock broken and removed. Over part of the land no

[1] *Journal of the Royal Agricultural Society*, 1864, p. 369.

Fig. 8. Coxheath, Maidstone. Once waste land, now reclaimed and valuable farm land

further treatment seems to have been necessary, and good crops were at once obtained after a reasonable outlay on manure; so permanent was the improvement that the land still lets for £2 per acre per annum. Other of the land, however, was very poor and required heavy manuring before it became productive. Most of the land thus reclaimed was divided among the lords of the manors and others possessing rights of common, of cutting turf, etc., while part of the remainder was sold to defray the expense of reclamation; the stone that was taken out lay stacked along the roads in enormous quantities, and people thought it never could be used, but subsequently it was all required for making up the Weald roads. The reclamation went on between 1814 and 1818, but was not completed: only recently has the whole of the land been taken in, the last surviving piece of waste having proved a considerable nuisance because of the gipsies that encamped there.

The system of managing sandy soils introduced by Townshend and Coke is, as we have seen, a combination of crops and live stock: nitrogenous compounds are added to the soil by clover or other leguminous crops, and by purchased oilcake: lime, potassium salts, phosphates are also added: the crops so grown are (with the exception of grain) fed to animals to make manure for more crops. Crops and live stock are thus mutually interdependent, and any break-down on the part of either causes the collapse of the system.

Another method of dealing with sandy soils has, however, long been practised. The ease with which they are cultivated and the earliness at which their crops ripen marked them out long ago as eminently suitable for market garden produce. The light loams of East Kent have grown fruit since Tudor times. Vegetables and fruit were long ago grown on the light soils round London, and the practice was extending by the end of the eighteenth century even in places as remote as Suffolk; stable manure was barged from London to Manningtree and sold at 10s. for a five-horse load at the quay, while carrots were grown on the sandy soils and sent back to London. This system has now developed very extensively, and now-a-days considerable areas of sand produce potatoes, vegetables or fruit to be sent off to the cities, and are fertilised with stable manure and other refuse brought back from the cities.

A third system is in use. The introduction of artificial manures has enabled the sand farmer to be wholly independent either of live stock or of city manures for keeping up the fertility of his soil. The classical instance of this type of management is afforded by the Lupitz estate at Altmark, Saxony.

When Herr Schultz came into possession in 1855 the land was largely a barren heath, yielding crops only at considerable expense. He soon observed, however, that lupins grew well provided the rainfall was sufficient (the average fall was 27 inches), and also found that they enriched the soil in nitrogenous organic matter and fertilised the next crop. His method as finally worked out was essentially as follows: leguminous crops were grown and fertilised with mineral manures—lime, phosphates and potassium salts—in order to induce considerable plant development and therefore considerable nitrogen fixation. Some of the crops were cut for hay, but in the main they were ploughed in, thus adding to the soil the nitrogenous organic matter of the stems and leaves as well as of the roots. Crops thus ploughed in are called green manure. An acre of green crop was found to furnish roughly as much nitrogen as 10 to 15 tons of farmyard manure, and this, taken in conjunction with the nutrients added in the artificial fertilisers, sufficed to yield large crops of grain and potatoes. Thus, without purchasing nitrogenous manures (which are very costly), and without keeping much live stock, Schultz was able by periodical green manuring enormously to increase the productiveness of his land. This extremely valuable method has been much adopted in Germany.

Where high priced crops are grown it may prove

more profitable to purchase nitrogenous manure
instead of making it through the agency of leguminous crops. Dr C. S. Edwards adopted this course
in his recent successful reclamation of a considerable
area of derelict sand at Tangham, Capel St Andrew,
Suffolk. The heather and bracken having been
eradicated, 20 to 25 loads per acre of "crag" (finely
divided shells, silts, etc., rich in calcium carbonate
and occurring in deposits just below the surface)
were put on at a cost of about £1 to £2 according
to the distance from the pit, and then the land was
broken up. It was usually necessary to do this with
the steam cultivator because the large number of
rabbit holes made the ground dangerous for horses.
After about four to eight cultivations the land was
sufficiently level, the rabbit holes filled and the weeds
killed: this cost up to £2 per acre. A second cragging
is now advantageous. The crops grown are wheat,
oats, potatoes, blue peas (sold dried for human consumption) and carrots; for the working horses and
the pigs a patch of lucerne is grown. As everything
is sold off and no stock is kept (except pigs to
consume waste potatoes, etc.), it is unnecessary to
adhere to any rigid rotation, and the farmer can
grow whatever promises to yield most profit. The
artificial manures used are: for carrots and potatoes,
1 cwt. of muriate of potash and 1½ cwt. of bone meal
applied in April, 1½ cwt. *each* of nitrate of soda or

lime and sulphate of ammonia applied in two dressings later on; for wheat or oats, 1 cwt. nitrate of soda or lime or sulphate of ammonia, $\frac{1}{2}$ cwt. sulphate of potash and $\frac{3}{4}$ cwt. of bone meal; for peas, $\frac{1}{2}$ cwt. sulphate of potash and $\frac{3}{4}$ cwt. bone meal. Such little dung as is got goes on the lucerne or on the wheat. Careful cultivation is necessary to preserve a fine mulch, whereby the soil retains its moisture, and also to keep down weeds which are apt to be a considerable nuisance: even the freshly-broken ground covered itself with spear grass (couch), makebeg (spurry) and sorrel, while later on iron weed (polygonum), pansy, cranesbill and others came in.

The great advantage of the system is that it can be worked with but little capital and at a minimum of risk. It is therefore well adapted for small holdings, for which purpose, indeed, it was devised. The returns have been very satisfactory: the sales have averaged £6 to £7 per acre, while a rent of 15s. per acre was found to pay 5 per cent. on the cost of reclamation, 5 per cent. on a sinking fund, and also the rates and the rental imposed by the superior landlord.

The occupier of a sandy soil has therefore several possibilities open to him. If he can command capital he can go in for live stock and work on the Townshend-Coke system. If he prefers cultivation he can go in for intensive market gardening. If he has little or no capital he can work on Dr Edward's lines and grow the more highly priced of the ordinary crops by the use of suitable artificial manures.

An instance may be given of the modern development of the Townshend-Coke treatment of light sandy soils. The farm is on a light sand in Surrey, and is so dry in summer that satisfactory grass cannot be grown; no sheep are therefore kept during the warm months of the year. Sheep are bought in from Sussex and the West Country in September and are fattened on the land during the winter; they are sold as fast as they are ready and are all cleared out by the end of May. Considerable amounts of green crops are grown for them, including trifolium, green rye, rape, kale, turnips, swedes, and a mixture of clover and rye-grass, whilst large quantities of oil-cake and purchased grains are also supplied. The land thus becomes well fertilised, and is now sown with spring oats, which are often succeeded by a crop of malting barley. Then a mixture of rye-grass and clover is sown to afford hay in June and green food in September. The land is then ploughed up for winter wheat, and a dressing of London stable manure is given so as to ensure a satisfactory crop of straw which is sometimes a very saleable commodity. Recourse is had to artificial manures and periodically to lime in order to maintain fertility at a high level.

Instances of the market garden method occur on

almost any sandy soil within access of a large city,
especially where one man happens to have prospered
and so given the locality a reputation. No general
rules can be given about the management : the
successful grower generally keeps his land continu-
ously cropped and carefully watches the markets so
as to grow those things likely to yield most profit.
The following is an actual example. The grower is
near to London and has access to a riverside wharf;
he buys as manure City refuse, cleanings from
cattle steamers, unsold lots of stable manure, con-
demned fish, and any odds and ends of manurial
value. He also takes for a consideration some of
the local sewage. His ground is never idle: early
potatoes, onions, sprouting broccoli, peas and other
crops succeed each other without delay, odd corners
are filled up with early carrots, radishes, etc., all the
crops are carefully nursed so as to be ready for market
while prices are still high, i.e. before other people
have their produce ready. Much fruit is grown, pigs
are kept in the orchards to do the cultivation and
devour unsaleable crops. Considerable advantage is
taken of labour-saving devices. The grower's success
does not stop at production, but extends also to the
business side. Market garden produce is sold to
salesmen in the large markets who in turn have to
keep contracts with large customers. This grower,
being eminently successful and dependable even in

bad seasons, is therefore a useful stand-by for these people, and receives respectful consideration at their hands.

Considerable areas of sand, however, still lie waste in the agricultural sense. But they are not necessarily unproductive from other points of view. Some of the most delightful scenery in our country is to be found on the sands; even the most ardent reformer would hardly wish to root up the New Forest, the Bournemouth pine woods, the Wareham and Dorchester heaths, and substitute fields of turnips and sheep. Elsewhere, also, golf links have proved extremely remunerative. Considerable tracts of sand are given up to game. And there lies one of the difficulties of the situation. For game sometimes plays a very large part in the economy of the countryside, and may dominate pretty completely the movements both of man and of beast. There are parts of East Suffolk where the cottagers' cats have to live chained up like yard dogs in order to be safe from the gamekeepers' guns, while extensive damage is often done to the crops by birds, hares and rabbits. And so it happens that the man who undertakes to reclaim derelict sand and bring it into cultivation has not only to overcome the natural difficulties of the problem, but also to come to terms with the game preservers.

Lastly, the sand dunes are now beginning to

attract attention. It has long been the practice to plant these up with conifers or other suitable trees, but in New Zealand dune pastures are produced instead. Marram grass (*Ammophila arundinacea*) is first grown to fix the sand, and then the tree lupins (*Lupinus arboreus*); finally pasture grasses are sown, especially such as, under the prevailing conditions, make considerable root and so add to the stores of soil organic matter.

CHAPTER VIII

THE MOOR—WHAT SHALL IT BECOME?

WE have already dealt with the reclamation of the fens, we must now turn to the wholly distinct case of the moors. Speaking generally the moorland tracts of the country lie in high regions of considerable rainfall : the fens, on the other hand, are low lying and have a much smaller rainfall. When the fens are drained they become at once fit for cultivation and yield considerable quantities of potatoes, wheat and other crops. But the high moors do not : their soil is fundamentally different ; their rainfall is too high for satisfactory crop production ; and owing to their great altitude the winters are so severe that any kind of farming is attended with risk and difficulty.

Low-lying moor and peat districts may escape these climatic disadvantages and then their reclamation becomes simply a matter of soil treatment. Chat Moss in Lancashire affords an instance : it was first drained to remove excess of water and then heavily treated with town refuse from Manchester which enabled it to yield crops. An arrangement of this sort is mutually advantageous to city and country so that the expense can be distributed. But elsewhere the carriage of bulky extraneous matter becomes too costly to be borne entirely by the reclaimed land, and more concentrated ameliorating agents become necessary.

The problem has attracted considerable attention in Germany and Sweden and is under investigation at the special experiment stations at Bremen and Jönköping. Enough has been done to show that treatment with artificial manures leads to profitable crop production, especially of rye, oats and potatoes, on drained land. These crops do not stand in great want of lime, and need only a few hundredweights per acre of potassic, phosphatic and nitrogenous fertilisers. But when it is desired to diversify the agriculture, liming becomes necessary and may prove costly. In some districts Hiltner has successfully solved the problem of growing leguminous crops, and has thus opened up the possibility of raising potatoes and cereals more cheaply than before.

An interesting piece of reclamation went on under Nilsson's direction in Gothland, Sweden. More than 30,000 hectares of the island consisted of barren swamps, and yet the soil was rich both in lime and in nitrogen. Nilsson showed that the limiting factor was lack of phosphorus. The ground was therefore drained and treated with phosphatic manure ; it then yielded excellent crops of corn, rape and sugar beet. Instead of depending wholly on foreign supplies of phosphates, investigation was made by Wiborgh of the iron ore occurring in Northern Sweden and known to contain calcium phosphate : a method of extraction was devised and quantities of this Wiborgh phosphate were produced at Luliå. Subsequently a cheaper process was worked out by Palmaer.

In the Isle of Ely phosphates are also found to be the limiting factor and dressings of superphosphate result in marked crop production ; elsewhere, however, potassium compounds constitute the limiting factor. A well-known example is furnished by the Momence experimental field set out by Dr Cyril Hopkins in Illinois, where potassic fertilisers yield good crops but other fertilisers prove useless. Dr Hopkins[1] tells a pathetic story of a settler who spent years of unavailing labour on some of this black soil and did not

[1] *The Story of the Soil*, Boston, 1912.

find out till too late the one thing needful. The man had brought his wife and children to see the heavy crop on plots treated with potassic fertiliser alongside of the miserable one on the untreated land. "As he stood looking, first at the corn on the treated and untreated land, and then at his wife and large family of children, he broke down and cried like a child. Later he explained to the superintendent who was showing him the experiments, that he had put the best of his life into that kind of land. 'The land looked rich,' said he—'as rich as any land I ever saw. I bought it and drained it and built my home on a sandy knoll. The first crops were fairly good, and we hoped for better crops; but instead they grew worse and worse. We raised what we could on a small patch of sandy land, and kept trying to find out what we could grow on this black bogus land. Sometimes I helped the neighbours and got a little money, but my wife and I and my older children have wasted twenty years on this land. Poverty, poverty always. How was I to know that this single substance which you call potassium was all we needed to make this land productive and valuable?'"

The peats can be made productive if the climatic conditions are favourable. The agricultural utilisation of moorland depends mainly on this consideration. Obviously nothing but actual experiment would determine what could be accomplished, but at the present moment there seems little likelihood of any serious attempt being made in this country. For game has taken possession of the moorland, as of other land of low agricultural value, and having once got possession it is not easily displaced. A system of farming is in vogue that does not clash with game ; sheep are allowed to graze on the moorland, while the lower fields are kept in grass to furnish a little hay when needed. The land thus yields two rents : the shooting tenant on a Lancashire moor may pay 3s. 6d. per acre, while the agricultural tenant pays 2s., a total of 5s. 6d. Any agricultural development that involved displacement of the game would have to yield a sufficiently increased rent over the whole moor or the owner would lose financially, and it is difficult to see on our present knowledge what agricultural system could do this[1]. Only a properly conducted experiment could decide the matter.

[1] An instance is given by Mr Pell in the *Journal of the Royal Agricultural Society* for 1887, where £24. 7s. 6d. was spent per acre in reclaiming a moor, and the increased rental only amounted to 3s. 2d. per acre.

CHAPTER IX

CONCLUSION

WE can now sum up the general conclusions to which the previous chapters lead.

The problem of making a soil fertile consists in finding out first what conditions the plant requires for its proper development, and then altering the soil as far as possible to make it meet these requirements. If the discrepancy between the actual and the ideal is too great the plant may be altered by the breeder in such a way that its new requirements shall be nearer the actual possibilities of the case.

It is always most economical to select crops naturally adapted to the climatic and soil conditions, so that the gap to be bridged shall be small. Our problem is to alter the soil : in the first instance it is necessary to ascertain what the actual soil conditions are, and then to find which constitutes the limiting factor and to change that one. Probably another will now be found to set the limit : this must in turn be changed and so on until the limit is set by the incapacity of the plant to make further growth, and not by any soil factor. The responsibility of the soil investigator is now at an end, and the problem becomes one for the plant breeder.

The soil conditions fall mainly into four groups. (1) The physical constitution of the soil determines the movement of water, of air, and of the plant roots. (2) The chemical composition shows the amount of food materials present, and whether there are any detrimental substances in the soil. (3) The micro-organisms of the soil are very mixed, some of them work up certain of the food materials into forms suitable to the plant and are therefore eminently useful, others are detrimental in various ways. (4) Extrinsic factors such as climate, situation of the soil, nature of the subsoil, etc., play an important part. Any of these groups may determine the fertility of a particular soil. To a certain extent, however, all of them are under control. The physical constitution is altered and made more favourable to plant growth by adequate cultivation, by addition of organic matter (e.g. farmyard manure) and if necessary of lime, chalk, marl, or clay. The chemical composition may be entirely altered so far as food constituents are concerned by adding appropriate fertilisers. Control of the micro-organisms of the soil is as yet in its infancy although a beginning has been made. The extrinsic factors are naturally less easy to change, but the subsoil may be altered by breaking any pan or rock layer, and drainage may be effected by suitable means.

Soils do not fall into sharply defined classes but form a perfect gradation from intractable clays through the loams to light blowing sands. The loams are very fertile: the extremes of clay and of sand are infertile. But there is no sharp end-point: a number of soils near the limit may be productive under one system of management and not under another, and may be in cultivation during times of prosperity and derelict when prices fall.

In applying these general principles to any particular case a considerable amount of balancing of probabilities is necessary. Means taken to alter the physical condition of the soil may react on the micro-organisms, the chemical composition, etc., and *vice versa.* Above all, as the soil is cultivated for profit, economic considerations come in at every turn. Thus fertility problems are usually more complex than they appear at first sight, and require to be studied in the laboratory, the pot culture house and the field before they can be regarded as solved.

BIBLIOGRAPHY

HALL, A. D.   The Soil.   (Murray, 1911.)

—— The Book of the Rothamsted Experiments.   (Murray, 1905.)

HOPKINS, CYRIL G.   Soil Fertility and Permanent Agriculture.   (Boston, 1910.)

PROTHERO, ROWLAND E.   English Farming past and present.   (Longmans, 1912.)

RUSSELL, E. J.   Soil Conditions and Plant Growth.   (Longmans, 1912.)

Fig. 9. Pot culture house at Rothamsted, where experiments can be tried on a small scale before they are carried into the field. The wire covering is to keep out birds

INDEX